Amos E. Dolbear

The Machinery of the Universe

Mechanical conceptions of physical phenomena

Amos E. Dolbear

The Machinery of the Universe
Mechanical conceptions of physical phenomena

ISBN/EAN: 9783337163907

Printed in Europe, USA, Canada, Australia, Japan

Cover: Foto ©berggeist007 / pixelio.de

More available books at **www.hansebooks.com**

THE MACHINERY OF THE UNIVERSE

MECHANICAL CONCEPTIONS OF PHYSICAL PHENOMENA

BY

A. E. DOLBEAR, A.B., A.M., M.E., Ph.D.

PROFESSOR OF PHYSICS AND ASTRONOMY, TUFTS COLLEGE, MASS.

PUBLISHED UNDER GENERAL LITERATURE COMMITTEE.

LONDON:

SOCIETY FOR PROMOTING CHRISTIAN KNOWLEDGE,

NORTHUMBERLAND AVENUE, W.C.;

43, QUEEN VICTORIA STREET, E.C.

BRIGHTON : 129, NORTH STREET.

NEW YORK: E. & J. B. YOUNG & CO.

1897.

PREFACE

FOR thirty years or more the expressions "Correlation of the Physical Forces" and "The Conservation of Energy" have been common, yet few persons have taken the necessary pains to think out clearly what mechanical changes take place when one form of energy is transformed into another.

Since Tyndall gave us his book called *Heat as a Mode of Motion* neither lecturers nor text-books have attempted to explain how all phenomena are the necessary outcome of the various forms of motion. In general, phenomena have been attributed to *forces*—a metaphysical term, which explains nothing and is merely a stop-gap, and is really not at all needful in these days, seeing that transformable modes of motion, easily perceived and understood, may be substituted in all cases for forces.

In December 1895 the author gave a lecture before the Franklin Institute of Philadelphia, on " Mechanical Conceptions of Electrical Phenomena," in which he undertook to make clear what happens when electrical phenomena appear. The publication of this lecture in *The Journal of the Franklin Institute* and in *Nature* brought an urgent request that it should be enlarged somewhat and published in a form more convenient for the public. The enlargement consists in the addition of a chapter on the " *Contrasted Properties of Matter and the Ether*," a chapter containing something which the author believes to be of philosophical importance in these days when electricity is so generally described as a phenomenon of the ether.

A. E. DOLBEAR.

TABLE OF CONTENTS

CHAPTER I

CHAPTER II

CHAPTER III

CHAPTER I

Ideas of phenomena ancient and modern, metaphysical and
mechanical—Imponderables—Forces, invented and dis-
carded—Explanations—Energy, its factors, Kinetic and
Potential—Motions, kinds and transformations of—
Mechanical, molecular, and atomic—Invention of Ethers,
Faraday's conceptions.

'And now we might add something concerning a most subtle
spirit which pervades and lies hid in all gross bodies, by
the force and action of which spirit the particles of bodies
attract each other at near distances, and cohere if con-
tiguous, and electric bodies operate at greater distances,
as well repelling as attracting neighbouring corpuscles,
and light is emitted, reflected, inflected, and heats bodies,
and all sensation is excited, and members of animal bodies
move at the command of the will.'—NEWTON, *Principia*.

IN Newton's day the whole field of nature was
practically lying fallow. No fundamental principles were known until the law of gravitation was
discovered. This law was behind all the work of
Copernicus, Kepler, and Galileo, and what they had
done needed interpretation. It was quite natural

7

that the most obvious and mechanical phenomena should first be reduced, and so the *Principia* was concerned with mechanical principles applied to astronomical problems. To us, who have grown up familiar with the principles and conceptions underlying them, all varieties of mechanical phenomena seem so obvious, that it is difficult for us to understand how any one could be obtuse to them ; but the records of Newton's time, and immediately after this, show that they were not so easy of apprehension. It may be remembered that they were not adopted in France till long after Newton's day. In spite of what is thought to be reasonable, it really requires something more than complete demonstration to convince most of us of the truth of an idea, should the truth happen to be of a kind not familiar, or should it chance to be opposed to our more or less well-defined notions of what it is or ought to be. If those who labour for and attain what they think to be the truth about any matter, were a little better informed concerning mental processes and the conditions under which ideas grow and displace others, they would be more patient with mankind ; teachers of every rank might then discover that what is often called stupidity may be nothing else than mental inertia, which can no more be made active by simply willing than can the movement of a cannon ball

by a like effort. We *grow* into our beliefs and
opinions upon all matters, and scientific ideas are
no exceptions.

Whewell, in his *History of the Inductive Sciences*,
says that the Greeks made no headway in physical
science because they lacked appropriate ideas.
The evidence is overwhelming that they were as
observing, as acute, as reasonable as any who live
to-day. With this view, it would appear that the
great discoverers must have been men who started
out with appropriate ideas : were looking for what
they found. If, then, one reflects upon the exceed-
ing great difficulty there is in discovering one new
truth, and the immense amount of work needed to
disentangle it, it would appear as if even the most
successful have but indistinct ideas of what is really
appropriate, and that their mechanical conceptions
become clarified by doing their work. This is not
always the fact. In the statement of Newton
quoted at the head of this chapter, he speaks of
a spirit which lies hid in all gross bodies, etc.,
by means of which all kinds of phenomena are
to be explained; but he deliberately abandons that
idea when he comes to the study of light, for
he assumes the existence and activity of light
corpuscles, for which he has no experimental
evidence; and the probability is that he did this
because the latter conception was one which he

could handle mathematically, while he saw no way
for thus dealing with the other. His mechanical
instincts were more to be trusted than his carefully
calculated results; for, as all know, what he called
"spirits," is what to-day we call the ether, and the
corpuscular theory of light has now no more than
a historic interest. The corpuscular theory was a
mechanical conception, but each such corpuscle
was ideally endowed with qualities which were out
of all relation with the ordinary matter with which
it was classed.

Until the middle of the present century the
reigning physical philosophy held to the existence
of what were called imponderables. The phenomena
of heat were explained as due to an imponderable
substance called "caloric," which ordinary matter
could absorb and emit. A hot body was one
which had absorbed an imponderable substance.
It was, therefore, no heavier than before, but it
possessed ability to do work proportional to the
amount absorbed. Carnot's ideal engine was described by him in terms that imply the materiality
of heat. Light was another imponderable substance, the existence of which was maintained by
Sir David Brewster as long as he lived. Electricity
and magnetism were imponderable fluids, which,
when allied with ordinary matter, endowed the
latter with their peculiar qualities. The con-

ceptions in each case were properly mechanical ones *part* (but not all) *of the time;* for when the immaterial substances were dissociated from matter, where they had manifested themselves, no one concerned himself to inquire as to their whereabouts. They were simply off duty, but could be summoned, like the genii in the story of Aladdin's Lamp. Now, a mechanical conception of any phenomenon, or a mechanical explanation of any kind of action, must be mechanical all the time, in the antecedents as well as the consequents. Nothing else will do except a miracle.

During the fifty years, from about 1820 to 1870, a somewhat different kind of explanation of physical events grew up. The interest that was aroused by the discoveries in all the fields of physical science— in heat, electricity, magnetism and chemistry—by Faraday, Joule, Helmholtz, and others, compelled a change of conceptions ; for it was noticed that each special kind of phenomenon was preceded by some other definite and known kind; as, for instance, that chemical action preceded electrical currents, that mechanical or electrical activity resulted from changing magnetism, and so on. As each kind of action was believed to be due to a special force, there were invented such terms as mechanical force, electrical force, magnetic, chemical and vital forces, and these were discovered to be

convertible into one another, and the "doctrine of
the correlation of the physical forces" became a
common expression in philosophies of all sorts.
By "convertible into one another," was meant,
that whenever any given force appeared, it was at
the expense of some other force ; thus, in a battery
chemical force was changed into electrical force ;
in a magnet, electrical force was changed into magnetic force, and so on. The idea here was the
transformation of forces, and *forces* were not so
clearly defined that one could have a mechanical
idea of just what had happened. That part of the
philosophy was no clearer than that of the imponderables, which had largely dropped out of mind.
The terminology represented an advance in knowledge, but was lacking in lucidity, for no one knew
what a force of any kind was.

The first to discover this and to repudiate the
prevailing terminology were the physiologists, who
early announced their disbelief in a vital force, and
their belief that all physiological activities were of
purely physical and chemical origin, and that there
was no need to assume any such thing as a vital
force. Then came the discovery that chemical force,
or affinity, had only an adventitious existence, and
that, at absolute zero, there was no such activity.
The discovery of, or rather the appreciation of,
what is implied by the term *absolute zero*, and

especially of the nature of heat itself, as expressed in the statement that heat is a mode of motion, dismissed another of the so-called forces as being a metaphysical agency having no real existence, though standing for phenomena needing further attention and explanation; and by explanation is meant *the presentation of the mechanical antecedents for a phenomenon, in so complete a way that no supplementary or unknown factors are necessary.* The train moves because the engine pulls it; the engine pulls because the steam pushes it. There is no more necessity for assuming a steam force between the steam and the engine, than for assuming an engine force between the engine and the train. All the processes are mechanical, and have to do only with ordinary matter and its conditions, from the coal-pile to the moving freight, though there are many transformations of the forms of motion and of energy between the two extremes.

During the past thirty years there has come into common use another term, unknown in any technical sense before that time, namely, *energy.* What was once called the conservation of force is now called the conservation of energy, and we now often hear of forms of energy. Thus, heat is said to be a form of energy, and the forms of energy are convertible into one another, as the so-called forces were formerly supposed to be transformable into one another.

We are asked to consider gravitative energy, heat energy, mechanical energy, chemical energy, and electrical energy.   When we inquire what is meant by energy, we are informed that it means ability to do work, and that work is measureable as a pressure into a distance, and is specified as foot-pounds.   A mass of matter moves because energy has been spent upon it, and has acquired energy equal to the work done on it, and this is believed to hold true, no matter what the kind of energy was that moved it. If a body moves, it moves because another body has exerted pressure upon it, and its energy is called *kinetic energy;* but a body may be subject to pressure and not move appreciably, and then the body is said to possess potential energy.   Thus, a bent spring and a raised weight are said to possess potential energy.   In either case, *an energized body receives its energy by pressure, and has ability to produce pressure on another body.*   Whether or not it does work on another body depends on the rigidity of the body it acts upon.   In any case, it is simply a mechanical action—body A pushes upon body B (Fig. 1).   There is no need to assume anything more mysterious than mechanical action.   Whether body B moves this way or that depends upon the direction of the push, the point of its application. Whether the body be a mass as large as the earth or as small as a molecule, makes no difference in

that particular. Suppose, then, that *a* (Fig. 2)
spends its energy on *b*, *b* on *c*, *c* on *d*, and so on.
The energy of *a* gives translatory motion to *b*, *b*

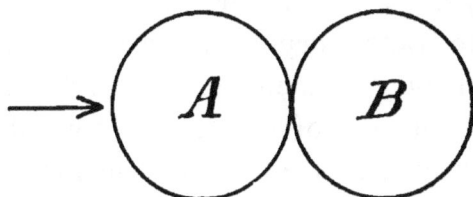

FIG. 1.

sets *c* vibrating, and *c* makes *d* spin on some axis.
Each of these has had energy spent on it, and
each has some form of energy different from the
other, but no new factor has been introduced
between *a* and *d*, and the only factor that has
gone from *a* to *d* has been motion—motion that

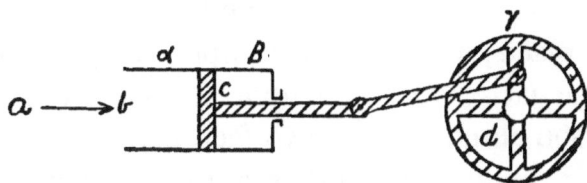

FIG. 2.

has had its direction and quality changed, but not
its nature. If we agree that energy is neither
created nor annihilated by any physical process,
and if we assume that *a* gave to *b* all its energy,
that is, all its motion ; that *b* likewise gave its all
to *c*, and so on ; then the succession of phenomena

from *a* to *d* has been simply the transference of a definite amount of motion, and therefore of energy, from the one to the other ; for *motion has been the only variable factor.* If, furthermore, we should agree to call the translatory motion *a*, the vibratory motion β, the rotary γ, then we should have had a conversion of *a* into β, of β into γ. If we should consider the amount of transfer motion instead of the kind of motion, we should have to say that the *a* energy had been transformed into β and the β into γ.

What a given amount of energy will do depends only upon its *form*, that is, the kind of motion that embodies it.

The energy spent upon a stone thrown into the air, giving it translatory motion, would, if spent upon a tuning fork, make it sound, but not move it from its place ; while if spent upon a top, would enable the latter to stand upon its point as easily as a person stands on his two feet, and to do other surprising things, which otherwise it could not do. One can, without difficulty, form a mechanical conception of the whole series without assuming imponderables, or fluids or forces. Mechanical motion only, by pressure, has been transferred in certain directions at certain rates. Suppose now that some one should suddenly come upon a spinning top (Fig. 3) while it was standing upon its point,

and, as its motion might not be visible, should
cautiously touch it.   It would bound away with surprising promptness, and, if he were not instructed in
the mechanical principles involved, he might fairly
well draw the conclusion that it was actuated by
other than simple mechanical principles, and, for
that reason, it would be difficult to persuade him

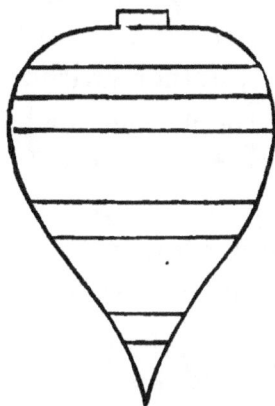

FIG. 3.

that there was nothing essentially different in the
body that appeared and acted thus, than in a stone
thrown into the air ; nevertheless, that statement
would be the simple truth.

   All our experience, without a single excep-
tion, enforces the proposition that no body moves
in any direction, or in any way, except when some
other body *in contact* with it presses upon it.   The
action is direct.   In Newton's letter to his friend

B

Bentley, he says—"That one body should act upon another through empty space, without the mediation of anything else by and through which their action and pressure may be conveyed from one to another, is to me so great an absurdity that I believe no man who has in philosophical matters a competent faculty of thinking can ever fall into it."

For mathematical purposes, it has sometimes been convenient to treat a problem as if one body could act upon another without any physical medium between them ; but such a conception has no degree of rationality, and I know of no one who believes in it as a fact. If this be granted, then our philosophy agrees with our experience, and every body moves because it is pushed, and the mechanical antecedent of every kind of phenomenon is to be looked for in some adjacent body possessing energy—that is, the ability to push or produce pressure.

It must not be forgotten that energy is not a simple factor, but is always a product of two factors—a mass with a velocity, a mass with a temperature, a quantity of electricity into a pressure, and so on. One may sometimes meet the statement that matter and energy are the two realities ; both are spoken of as entities. It is much more philosophical to speak of matter and motion, for in the absence of motion there is no energy, and the

energy varies with the amount of motion; and
furthermore, to understand any manifestation of
energy one must inquire what kind of motion is
involved. This we do when we speak of mechanical energy as the energy involved in a body having
a translatory motion ; also, when we speak of heat
as a vibratory, and of light as a wave motion. To
speak of energy without stating or implying these
distinctions, is to speak loosely and to keep far
within the bounds of actual knowledge. To speak
thus of a body possessing energy, or expending
energy, is to imply that the body possesses some
kind of motion, and produces pressure upon another
body because it has motion. Tait and others have
pointed out the fact, that what is called potential
energy must, in its nature, be kinetic. Tait says—
"Now it is impossible to conceive of a truly dor-
mant form of energy, whose magnitude should
depend, in any way, upon the unit of time ; and
we are forced to conclude that potential energy,
like kinetic energy, depends (even if unexplained
or unimagined) upon motion." All this means
that it is now too late to stop with energy as a
final factor in any phenomenon, that the *form of
motion* which embodies the energy is the factor
that determines *what* happens, as distinguished
from how *much* happens. Here, then, are to be
found the distinctions which have heretofore been

called forces ; here is embodied the proof that
direct pressure of one body upon another is what
causes the latter to move, and that the direction
of movement depends on the point of application,
with reference to the centre of mass.

It is needful now to look at the other term in
the product we call energy, namely, the substance
moving, sometimes called matter or mass. It has
been mentioned that the idea of a medium filling
space was present to Newton, but his gravitation
problem did not require that he should consider
other factors than masses and distances. The law
of gravitation as considered by him was—Every
particle of matter attracts every other particle of
matter with a stress which is proportional to the
product of their masses, and inversely to the
squares of the distance between them. 'Here we
are concerned only with the statement that every
particle of matter attracts every other particle of
matter. Everything then that possesses gravitative
attraction is matter in the sense in which that term
is used in this law. If there be any other substance in the universe that is not thus subject to
gravitation, then it is improper to call it matter,
otherwise the law should read, " Some particles of
matter attract," etc., which will never do.

We are now assured that there is something else
in the universe which has no gravitative property

at all, namely, the ether. It was first imagined in order to account for the phenomena of light, which was observed to take about eight minutes to come from the sun to the earth. Then Young applied the wave theory to the explanation of polarization and other phenomena; and in 1851 Foucault proved experimentally that the velocity of light was less in water than in air, as it should be if the wave theory be true, and this has been considered a crucial experiment which took away the last hope for the corpuscular theory, and demonstrated the existence of the ether as a space-filling medium capable of transmitting light-waves known to have a velocity of 186,000 miles per second. It was called the luminiferous ether, to distinguish it from other ethers which had also been imagined, such as electric ether for electrical phenomena, magnetic ether for magnetic phenomena, and so on—as many ethers, in fact, as there were different kinds of phenomena to be explained.

It was Faraday who put a stop to the invention of ethers, by suggesting that the so-called luminiferous ether might be the one concerned in all the different phenomena, and who pointed out that the arrangement of iron filings about a magnet was indicative of the direction of the stresses in the ether. This suggestion did not meet the approval of the mathematical physicists of his day, for it necessitated the abandonment of the conceptions they had worked with, as well as the terminology which had been employed, and made it needful to reconstruct all their work to make it intelligible —a labour which was the more distasteful as it was forced upon them by one who, although expert enough in experimentation, was not a mathematician, and who boasted that the most complicated mathematical work he ever did was to turn the crank of a calculating machine ; who did all his work, formed his conclusions, and then said— " The work is done ; hand it over to the computers."

It has turned out that Faraday's mechanical conceptions were right. Every one now knows of Maxwell's work, which was to start with Faraday's conceptions as to magnetic phenomena, and follow them out to their logical conclusions, applying them to molecules and the reactions of the latter upon the ether. Thus he was led to conclude that light was an electro-magnetic phenomenon ; that is, that the waves which constitute light, and the waves produced by changing magnetism were identical in their nature, were in the same medium, travelled with the same velocity, were capable of refraction, and so on. Now that all this is a matter of common knowledge to-day, it is curious to look back no further than ten years. Maxwell's conclusions

were adopted by scarcely a physicist in the world.
Although it was known that inductive action
travelled with finite velocity in space, and that an
electro-magnet would affect the space about it
practically inversely as the square of the distance,
and that such phenomena as are involved in tele-
phonic induction between circuits could have no
other meaning than the one assigned by Maxwell,
yet nearly all the physicists failed to form the only
conception of it that was possible, and waited for
Hertz to devise apparatus for producing interference before they grasped it. It was even then so
new, to some, that it was proclaimed to be a demonstration of the existence of the ether itself, as
well as a method of producing waves short enough
to enable one to notice interference phenomena.
It is obvious that Hertz himself must have had the
mechanics of wave-motion plainly in mind, or he
would not have planned such experiments. The
outcome of it all is, that we now have experimental
demonstration, as well as theoretical reason for
believing, that the ether, once considered as only
luminiferous, is concerned in all electric and magnetic phenomena, and that waves set up in it by
electro-magnetic actions are capable of being
reflected, refracted, polarized, and twisted, in the
same way as ordinary light-waves can be, and that
the laws of optics are applicable to both.

CHAPTER II

PROPERTIES OF MATTER AND ETHER

Properties of Matter and Ether compared—Discontinuity *versus* Continuity—Size of atoms—Astronomical distances —Number of atoms in the universe—Ether unlimited— Kinds of Matter, permanent qualities of—Atomic structure ; vortex-rings, their properties—Ether structureless —Matter gravitative, Ether not—Friction in Matter, Ether frictionless—Chemical properties—Energy in Matter and in Ether—Matter as a transformer of Energy—Elasticity —Vibratory rates and waves—Density—Heat—Indestructibility of Matter—Inertia in Matter and in Ether— Matter not inert—Magnetism and Ether waves—States of Matter—Cohesion and chemism affected by temperature —Shearing stress in Solids and in Ether—Ether pressure —Sensation dependent upon Matter—Nervous system not affected by Ether states—Other stresses in Ether— Transformations of Motion—Terminology.

A COMMON conception of the ether has been that it is a finer-grained substance than ordinary matter, but otherwise so like the latter that the laws found to hold good with matter were equally applicable to the ether, and hence the mechanical conceptions

24

formed from experience in regard to the one have
been transferred to the other, and the properties
belonging to one, such as density, elasticity, etc.,
have been asserted as properties of the other.

There is so considerable a body of knowledge
bearing upon the similarities and dissimilarities of
these two entities that it will be well to compare
them. After such comparison one will be better
able to judge of the propriety of assuming them
to be subject to identical laws.

I. MATTER IS DISCONTINUOUS.

Matter is made up of atoms having dimensions
approximately determined to be in the neighbourhood of the one fifty-millionth of an inch in
diameter. These atoms may have various degrees
of aggregation;—they may be in practical contact, as in most solid bodies such as metals and
rocks ; in molecular groupings as in water, and
in gases such as hydrogen, oxygen, and so forth,
where two, three, or more atoms cohere so strongly
as to enable the molecules to act under ordinary
circumstances like simple particles. Any or all
of these molecules and atoms may be separated by
any assignable distance from each other. Thus, in ·
common air the molecules, though rapidly changing
their positions, are on the average about two
hundred and fifty times their own diameter apart.

This is a distance relatively greater than the distance apart of the earth and the moon, for two hundred and fifty times the diameter of the earth will be $8000 \times 250 = 2,000,000$ miles, while the distance to the moon is but 240,000 miles. The sun is 93,000,000 miles from the earth, and the most of the bodies of the solar system are still more widely separated, Neptune being nearly 3000 millions of miles from the sun. As for the fixed stars, they are so far separated from us that, at the present rate of motion of the solar system in its drift through space—500 millions of miles in a year—it would take not less than 40,000 years to reach the nearest star among its neighbours, while for the more remote ones millions of years must be reckoned. The huge space separating these masses is practically devoid of matter; it is a vacuum.

THE ETHER IS CONTINUOUS.

The idea of continuity as distinguished from discontinuity may be gained by considering what would be made visible by magnification. Water appears to the eye as if it were without pores, but if sugar or salt be put into it, either will be dissolved and quite disappear among the molecules of the water as steam does in the air, which shows that there are some unoccupied spaces between the molecules.

If a microscope be employed to magnify a minute drop of water it still shows the same lack of structure as that looked at with the unaided eye. If the magnifying power be the highest it may reveal a speck as small as the hundred-thousandth part of an inch, yet the speck looks no different in character. We know that water is composed of two different kinds of atoms, hydrogen and oxygen, for they can be separated by chemical means and kept in separate bottles, and again made to combine to form water having all the qualities that belonged to it before it was decomposed. If a very much higher magnifying power were available, we should ultimately be able to see the individual water molecules, and recognize their hydrogen and oxygen constituents by their difference in size, rate of movements, and we might possibly separate them by mechanical methods. What one would see would be something very different in structure from the water as it appears to our eyes. If the ether were similarly to be examined through higher and still higher magnifying powers, even up to infinity, there is no reason for thinking that the last examination would show anything different in structure or quality from that which was examined with low power or with no microscope at all. This is all expressed by saying that the ether is a continuous substance, without interstices, that it fills space completely, and, unlike gases, liquids, and solids, is incapable of absorbing or dissolving anything.

2. MATTER IS LIMITED.

There appears to be a definite amount of matter in the visible universe, a definite number of molecules and atoms. How many molecules there are in a cubic inch of air under ordinary pressure has been determined, and is represented approximately by a huge number, something like a thousand million million millions.

When the diameter of a molecule has been measured, as it has been approximately, and found to be about one fifty-millionth of an inch, then fifty million in a row would reach an inch, and the cube of fifty million is 125,000,000000,000000,000000, one hundred and twenty-five thousand million million millions.   In a cubic foot there will of course be 1728 times that number.   One may if one likes find how many there may be in the earth, and moon, sun and planets, for the dimensions of them are all very well known.   Only the multiplication table need be used, and the sum of all these will give how many molecules there are in the solar system.   If one should feel that the number thus obtained was not very accurate, he might reflect that if there were ten times as many it would add but another cipher to a long line of similar ones and would not

materially modify it.   The point is that there is a
definite, computable number.   If one will then add
to these the number of molecules in the more
distant stars and nebulæ, of which there are visible
about 100,000,000, making such estimate of their
individual size as he thinks prudent, the sum of all
will give the number of molecules in the visible
universe.   The number is not so large but it can be
written down in a minute or two.   Those who have
been to the pains to do the sum say it may be
represented by seven followed by ninety-one ciphers.
One could easily compute how many molecules so
large a space would contain if it were full and as
closely packed as they are in a drop of water, but
there would be a finite and not an infinite number,
and therefore there is a limited number of atoms in
the visible universe.

THE ETHER IS UNLIMITED.

The evidence for this comes to us from the
phenomena of light.   Experimentally, ether waves
of all lengths are found to have a velocity of
186,000 miles in a second.   It takes about eight
minutes to reach us from the sun, four hours from
Neptune the most distant planet, and from the
nearest fixed star about three and a half years.
Astronomers tell us that some visible stars are so
distant that their light requires not less than ten

thousand years and probably more to reach us, though travelling at the enormous rate of 186,000 miles a second. This means that the whole of space is filled with this medium. If there were any vacant spaces, the light would fail to get through them, and stars beyond them would become invisible. There are no such vacant spaces, for any part of the heavens shows stars beaming continuously, and every increase in telescopic power shows stars still further removed than any seen before. The whole of this intervening space must therefore be filled with the ether. Some of the waves that reach us are not more than the hundred-thousandth of an inch long, so there can be no crack or break or absence of ether from so small a section as the hundred-thousandth of an inch in all this great expanse. More than this. No one can think that the remotest visible stars are upon the boundary of space, that if one could get to the most distant star he would have on one side the whole of space while the opposite side would be devoid of it. Space we know is of three dimensions, and a straight line may be prolonged in any direction to an infinite distance, and a ray of light may travel on for an infinite time and come to no end provided space be filled with ether.

How long the sun and stars have been shining no one knows, but it is highly probable that the sun has existed for not less than 1000 million years, and has during that time been pouring its rays as radiant energy into space.   If then in half that time, or 500 millions of years, the light had somewhere reached a boundary to the ether, it could not have gone beyond but would have been reflected back into the ether-filled space, and such part of the sky would be lit up by this reflected light.   There is no indication that anything like reflection comes to us from the sky.   This is equivalent to saying that the ether fills space in every direction away from us to an unlimited distance, and so far is itself unlimited.

3. MATTER IS HETEROGENEOUS.

The various kinds of matter we are acquainted with are commonly called the elements.   These when combined in various ways exhibit characteristic phenomena which depend upon the kinds of matter, the structure and motions which are involved.   There are some seventy different kinds of this elemental matter which may be identified as constituents of the earth.   Many of the same elements have been identified in the sun and stars, such for instance as hydrogen, carbon, and iron.   Such phenomena lead us to conclude that the kinds of matter elsewhere in the universe are identical with such as we are familiar with, and that elsewhere the variety is as great.   The qualities of the elements, within a certain range of temperature, are permanent; they are not subject to fluctuations, though the qualities of combinations of them may vary indefinitely. The elements therefore may be regarded as retaining their identity in all ordinary experience.

THE ETHER IS HOMOGENEOUS.

One part of the ether is precisely like any other part everywhere and always, and there are no such distinctions in it as correspond with the elemental forms of matter.

4. MATTER IS ATOMIC.

There is an ultimate particle of each one of the elements which is practically absolute and known as an atom. The atom retains its identity through all combinations and processes. It may be here or there, move fast or slow, but its atomic form persists.

THE ETHER IS NON-ATOMIC.

One might infer, from what has already been said about continuity, that the ether could not be constituted of separable particles like masses of matter; for no matter how minute they might be, there would be interspaces and unoccupied spaces which would present us with phenomena which have never been seen. It is the general consensus of opinion among those who have studied the subject that the ether is not atomic in structure.

5. MATTER HAS DEFINITE STRUCTURE.

Every atom of every element is so like every other atom of the same element as to exhibit the same characteristics, size, weight, chemical activity, vibratory rate, etc., and it is thus shown conclusively that the structural form of the elemental particles is the same for each element, for such characteristic reactions as they exhibit could hardly be if they were mechanically unlike.

Of what form the atoms of an element may be is not very definitely known. The earlier philosophers assumed them to be hard round particles, but later thinkers have concluded that atoms of such a character are highly improbable, for they could not exhibit in this case the properties which the elements do exhibit. They have therefore dismissed such a conception from consideration. In place of this hypothesis has been substituted a very different idea, namely, that an atom is a vortex-ring [1] of ether floating in the ether, as a smoke-ring

[1] Vortex-rings for illustration may be made by having a wooden box about a foot on a side, with a round orifice in the middle of one side, and the side opposite covered with stout cloth stretched tight over a framework. A saucer con-

puffed out by a locomotive in still air may float in
the air and show various phenomena.

A vortex-ring produced in the air behaves in the
most surprising manner.

FIG. 4.—Method of making vortex-rings and their behaviour.

taining strong ammonia water, and another containing strong
hydrochloric acid, will cause dense fumes in the box, and a
tap with the hand upon the cloth back will force out a ring
from the orifice. These may be made to follow and strike
each other, rebounding and vibrating, apparently attracting
each other and being attracted by neighbouring bodies.

By filling the mouth with smoke, and pursing the lips as if
to make the sound *o*, one may make fifteen or twenty small
rings by snapping the cheek with the finger.

1. It retains its ring form and the same material rotating as it starts with.

2. It can travel through the air easily twenty or thirty feet in a second without disruption.

3. Its line of motion when free is always at right angles to the plane of the ring.

4. It will not stand still unless compelled by some object. If stopped in the air it will start up itself to travel on without external help.

5. It possesses momentum and energy like a solid body.

6. It is capable of vibrating like an elastic body, making a definite number of such vibrations per second, the degree of elasticity depending upon the rate of vibration. The swifter the rotation, the more rigid and elastic it is.

7. It is capable of spinning on its own axis, and thus having rotary energy as well as translatory and vibratory.

8. It repels light bodies in front of it, and attracts into itself light bodies in its rear.

9. If projected along parallel with the top of a long table, it will fall upon it every time, just as a stone thrown horizontally will fall to the ground.

10. If two rings of the same size be travelling in the same line, and the rear one overtakes the other, the front one will enlarge its diameter, while the rear one will contract its own, till it can go through the forward one, when each will recover its original diameter, and continue on in the same direction, but vibrating, expanding and contracting their diameters with regularity.

11. If two rings be moving in the same line, but in opposite directions, they will repel each other when near, and thus retard their speed. If one goes through the other, as in the former case, it may quite lose its velocity, and come to a standstill in the air till the other has moved

on to a distance, when it will start up in its former direction.

12. If two rings be formed side by side, they will instantly collide at their edges, showing strong attraction.

13. If the collision does not destroy them, they may either break apart at the point of the collision, and then weld together into a single ring with twice the diameter, and then move on as if a single ring had been formed, or they may simply bounce away from each other, in which case they always rebound *in a plane* at right angles to the plane of collision. That is, if they collided on their sides, they would rebound so that one went up and the other down.

14. Three may in like manner collide and fuse into a single ring.

Such rings formed in air by a locomotive may rise wriggling in the air to the height of several hundred feet, but they are soon dissolved and disappear. This is because the friction and viscosity of the air robs the rings of their substance and energy. If the air were without friction this could not happen, and the rings would then be persistent, and would retain all their qualities.

Suppose then that such rings were produced in a medium without friction as the ether is believed to be, they would be permanent structures with a variety of properties. They would occupy space, have definite form and dimensions, momentum, energy, attraction and repulsion, elasticity; obey the laws of motion, and so far behave quite like such matter as we know. For such reasons it is thought by some persons to be not improbable that the atoms of matter are minute vortex-rings of ether in the ether. That which distinguishes the atom from the ether is the form of motion which is embodied in it, and if the motion were simply arrested, there would be nothing to distinguish the atom from the ether into which it dissolved. In other words, such a conception makes the atoms of matter a form of motion of the ether, and not a created something put into the ether.

THE ETHER IS STRUCTURELESS.

If the ether be the boundless substance described, it is clear it can have no form as a whole, and if it be continuous it can have no minute structure. If not constituted of atoms or molecules there is nothing descriptive that can be said about it. A molecule or a particular mass of matter could be identified by its form, and is thus in marked contrast with any portion of ether, for the latter could not be identified in a similar way. One may therefore say that the ether is formless.

6. MATTER IS GRAVITATIVE.

The law of gravitation is held as being universal. According to it every particle of matter in the universe attracts every other particle. The evidence for this law in the solar system is complete. Sun, planets, satellites, comets and meteors are all controlled by gravitation, and the movements of double stars testify to its activity among the more distant bodies of the universe. The attraction does not depend upon the kind of matter nor the arrangement of molecules or atoms, but upon the amount or mass of matter present, and if it be of a definite kind of matter, as of hydrogen or iron, the gravitative action is proportional to the number of atoms.

THE ETHER IS GRAVITATIONLESS.

One might infer already that if the ether were structureless, physical laws operative upon such material substances as atoms could not be applicable to it, and so indeed all the evidence we have shows that gravitation is not one of its properties. If it were, and it behaved in any degree like atomic structures, it would be found to be denser in the neighbourhood of large bodies like the earth, planets, and the sun. Light would be turned from its straight path while travelling in such denser medium, or made to move with less velocity. There is not the slightest indication of any such effect anywhere within the range of astronomical vision.

Gravitation then is a property belonging to

matter and not to ether. The impropriety of
thinking or speaking of the ether as matter of any
kind will be apparent if one reflects upon the
significance of the law of gravitation as stated.
Every particle of matter in the universe attracts
every other particle. If there be anything else in
the universe which has no such quality, then it
should not be called matter, else the law should
read : Some particles of matter attract some other
particles, which would be no law at all, for a real
physical law has no exceptions any more than the
multiplication table has. Physical laws are
physical relations, and all such relations are
quantitative.

7. MATTER IS FRICTIONABLE.

A bullet shot into the air has its velocity continuously reduced by the air, to which its energy is
imparted by making it move out of its way. A
railway train is brought to rest by the friction brake
upon the wheels. The translatory energy of the
train is transformed into the molecular energy called
heat. The steamship requires to propel it fast, a
large amount of coal for its engines, because the
water in which it moves offers great friction—
resistance which must be overcome. Whenever
one surface of matter is moved in contact with
another surface there is a resistance called friction,

the moving body loses its rate of motion, and will presently be brought to rest unless energy be continuously supplied. This is true for masses of matter of all sizes and with all kinds of motion. Friction is the condition for the transformation of all kinds of mechanical motions into heat. The test of the amount of friction is the rate of loss of motion. A top will spin some time in the air because its point is small. It will spin longer on a plate than on the carpet, and longer in a vacuum than in the air, for it does not have the air friction to resist it, and there is no kind or form of matter not subject to frictional resistance.

THE ETHER IS FRICTIONLESS.

The earth is a mass of matter moving in the ether. In the equatorial region the velocity of a point is more than a thousand miles in an hour, for the circumference of the earth is 25,000 miles, and it turns once on its axis in 24 hours, which is the length of the day. If the earth were thus spinning in the atmosphere, the latter not being in motion, the wind would blow with ten times hurricane velocity. The friction would be so great that nothing but the foundation rocks of the earth's crust could withstand it, and the velocity of rotation would be reduced appreciably in a relatively short time. The air moves along with the earth as a part of it, and consequently no such frictional destruction takes place, but the earth rotates in the ether with that same rate, and if the ether offered resistance it would react so as to retard the rotation and increase the length of the day. Astronomical observations show that the length of the day has certainly not changed 'so much as the tenth of a second during the past 2000 years. The earth also revolves about the sun, having a speed of about 19 miles in a second, or 68,000 miles an hour. This motion of the earth and the other planets about the sun is one of the most stable phenomena we know. The mean distance and period of revolution of every planet is unalterable in the long run. If the earth had been retarded by its friction in the ether the length of the year would have been changed, and astronomers would have discovered it. They assert that a change in the length of a year by so much as the hundredth part of a second has not happened during the past thousand years. This then is testimony, that a velocity of nineteen miles a second for a thousand years has produced no effect upon the earth's motion that is noticeable. Nineteen miles a second is not a very swift astronomical motion, for comets have been known to have a velocity of 400 miles a second when in the neighbourhood of the sun, and yet they have not

seemed to suffer any retardation, for their orbits
have not been shortened. Some years ago a comet
was noticed to have its periodic time shortened an
hour or two, and the explanation offered at first
was that the shortening was due to friction in the
ether although no other comet was thus affected.
The idea was soon abandoned, and to-day there
is no astronomical evidence that bodies having
translatory motion in the ether meet with any
frictional resistance whatever.   If a stone could be
thrown in interstellar space with a velocity of fifty
feet a second it would continue to move in a
straight line with the same speed for any assignable
time.

As has been said, light moves with the velocity
of 186,000 miles per second, and it may pursue its
course for tens of thousands of years.   There is no
evidence that it ever loses either its wave-length
or energy.   It is not transformed as friction would
transform it, else there would be some distance at
which light of given wave-length and amplitude
would be quite extinguished.   The light from
distant stars would be different in character from
that coming from nearer stars.   Furthermore, as
the whole solar system is drifting in space some
500,000,000 of miles in a year, new stars would be
coming into view in that direction, and faint stars
would be dropping out of sight in the opposite

direction — a phenomenon which has not been observed. Altogether the testimony seems conclusive that the ether is a frictionless medium, and does not transform mechanical motion into heat.

8. MATTER IS ÆOLOTROPIC.

That is, its properties are not alike in all directions. Chemical phenomena, crystallization, magnetic and electrical phenomena show each in their way that the properties of atoms are not alike on opposite faces. Atoms combine to form molecules, and molecules arrange themselves in certain definite geometric forms such as cubes, tetrahedra, hexagonal prisms and stellate forms, with properties emphasized on certain faces or ends. Thus quartz will twist a ray of light in one direction or the other, depending upon the arrangement which may be known by the external form of the crystal. Calc spar will break up a ray of light into two parts if the light be sent through it in certain directions, but not if in another. Tourmaline polarizes light sent through its sides and becomes positively electrified at one end while being heated. Some substances will conduct sound or light or heat or electricity better in one direction than in another. All matter is magnetic in some degree, and that implies polarity. If one will recall the structure of a vortex-ring, he will see how all the motion is inward on one side and outward on the other, which gives different properties to the two sides : a push away from it on one side and a pull toward it on the other.

THE ETHER IS ISOTROPIC.

That is, its properties are alike in every direction. There is no distinction due to position. A mass of matter will move as freely in one direction as in another ; a ray of light of any wave-length will travel in it in one direction as freely as in any other ; neither velocity nor direction are changed by the action of the ether alone.

9. MATTER IS CHEMICALLY SELECTIVE.

When the elements combine to form molecules they always combine in definite ways and in definite proportions. Carbon will combine with hydrogen, but will drop it if it can get oxygen. Oxygen will combine with iron or lead or sodium, but cannot be made to combine with fluorine. No more than two atoms of oxygen can be made to unite with one carbon atom, nor more than one hydrogen with one chlorine atom. There is thus an apparent choice for the kind and number of associates in molecular structure, and the instability of a molecule depends altogether upon the presence in its neighbourhood of other atoms for which some of the elements in the molecule have a stronger attraction or affinity than they have for the atoms they are now combined with. Thus iron is not stable in the presence of water molecules, and it becomes iron oxide; iron oxide is not stable in the presence of hot sulphur, it becomes an iron sulphide. All the elements are thus selective, and it is by such means that they may be chemically identified.

There is no phenomenon in the ether that is comparable with this. Evidently there could not be unless there were atomic structures having in some degree different characteristics which we know the ether to be without.

10. THE ELEMENTS OF MATTER ARE HARMONIC-ALLY RELATED.

It is possible to arrange the elements in the order of their atomic weights in columns which will show communities of property. Newlands, Mendeléeff, Meyer, and others have done this. The explanation for such an arrangement has not yet been forthcoming, but that it expresses a real fact is certain, for in the original scheme there were several gaps representing undiscovered elements, the properties of which were predicted from that of their associates in the table. Some of these have since been discovered, and their atomic weight and physical properties accord with those predicted.

With the ether such a scheme is quite impossible, for the very evident reason that there are no different things to have relation with each other. Every part is just like every other part. Where there are no differences and no distinctions there can be no relations. The ether is quite harmonic without relations.

11. MATTER EMBODIES ENERGY.

So long as the atoms of matter were regarded as hard round particles, they were assumed to be inert and only active when acted upon by what were called forces, which were held to be entities of some sort, independent of matter. These could pull or push it here or there, but the matter was itself incapable of independent activity. All this is now changed, and we are called upon to consider every atom as being itself a form of energy in the same sense as heat or light are forms of energy, the energy being embodied in particular forms of motion. Light, for instance, is a wave motion of the ether. An atom is a rotary ring of ether. Stop the wave motion, and the light would be annihilated. Stop the rotation, and the atom would be annihilated for the same reason. As the ray of light is a particular embodiment of energy, and has no existence apart from it, so an atom is to be regarded as an embodiment of energy. On a

previous page it is said that energy is the ability
of one body to act upon and move another in
some degree. An atom of any kind is not the
inert thing it has been supposed to be, for it can do
something. Even at absolute zero, when all its
vibratory or heat energy would be absent, it would
be still an elastic whirling body pulling upon
every other atom in the universe with gravitational
energy, twisting other atoms into conformity with
its own position with its magnetic energy ; and, if
such ether rings are like the rings which are made
in air, will not stand still in one place even if no
others act upon it, but will start at once by its own
inherent energy to move in a right line at right
angles to its own plane and in the direction of the
whirl inside the ring. Two rings of wood or iron
might remain in contact with each other for an
indefinite time, but vortex-rings will not, but will
beat each other away as two spinning tops will do
if they touch ever so gently. If they do not thus
separate it is because there are other forms of
energy acting to press them together, but such
external pressure will be lessened by the rings' own
reactions.

It is true that in a frictionless medium like the
ether one cannot at present see how such vortex-
rings could be produced in it. Certainly not by
any such mechanical methods as are employed to

make smoke-rings in air, for the friction of the air is the condition for producing them. However they came to be, there is implied the previous existence of the ether and of energy in some form capable of acting upon it in a manner radically different from any known in physical science.

There is good spectroscopic evidence that in some way elements of different kinds are now being formed in nebulæ, for the simplest show the presence of hydrogen alone. As they increase in complexity other elements are added, until the spectrum exhibits all the elements we know of. It has thus seemed likely either that most of what are called elements are composed of molecular group-ings of some fundamental element, which by proper physical methods might be decomposed, as one can now decompose a molecule of ammonia or sulphuric acid, or that the elements are now being created by some extra-physical process in those far-off regions. In either case an atom is the embodiment of energy in such a form as to be permanent under ordinary physical circumstances, but of which, if in any manner it should be destroyed, only the form would be lost. The ether would remain, and the energy which was embodied would be distributed in other ways.

THE ETHER IS ENDOWED WITH ENERGY.

The distinction between energy in matter and energy in the ether will be apparent, on considering that both the ether and energy in some form must be conceived as existing independent of matter; though every atom were annihilated, the ether would remain and all the energy embodied in the atoms would be still in existence in the ether. The atomic energy would simply be dissolved. One can easily conceive the ether as the same space-filling, continuous, unlimited medium, without an atom in it. On this assumption it is clear that no form of energy with which we have to deal in physical science would have any existence in the ether; for every one of those forms, gravitational, thermal, electric, magnetic, or any other — all are the results of the forms of energy in matter. If there were no atoms, there would be no gravitation, for that is the attraction of atoms upon each other. If there were no atoms, there could be no atomic vibration, therefore no heat, and so on for each and all. Nevertheless, if an atom be the embodiment of energy, there must have been energy in the ether before any atom existed. One of the properties of the ether is its ability to distribute energy in certain ways, but there is no evidence that of itself it ever transforms energy. Once a

D

given kind of energy is in it, it does not change ; hence for the apparition of a form of energy, like the first vortex-ring, there must have been not only energy, but some other agency capable of transforming that energy into a permanent structure. To the best of our knowledge to-day, the ether would be absolutely helpless. Such energy as was active in forming atoms must be called by another name than what is appropriate for such transformations as occur when, for instance, the mechanical energy of a bullet is transformed into heat when the target is struck. Behind the ether must be assumed some agency, directing and controlling energy in a manner totally different from any agency, which is operative in what we call physical science. Nothing short of what is called a miracle will do— an event without a physical antecedent in any way necessarily related to its factors, as is the fact of a stone related to gravity or heat to an electric current.

Ether energy is an endowment instead of being an embodiment, and implies antecedents of a super-physical kind.

12. MATTER IS AN ENERGY TRANSFORMER.

As each different kind of energy represents some specific form of motion, and *vice versâ,* some sort of mechanism is needful for transforming one kind into another, therefore molecular structure of one kind or another is essential. The transformation is a mechanical process, and matter in some particular and appropriate form is the condition of its taking place. If heat appears, then its antecedent has been some other form of motion acting upon the substance heated. It may have been the mechanical motion of another mass of matter, as when a bullet strikes a target and becomes heated; or it may be friction, as when a car-axle heats when run without proper oiling to reduce friction; or it may be condensation, as when tinder is ignited by condensing the air about it; or chemical reactions, when molecular structure is changed as in combustion, or an electrical current, which implies a dynamo and steam-engine or water-power. If light appears, its antecedent has been impact or friction, condensation or chemical action, and if electricity appears the same sort of antecedents are present. Whether the one or the other of these forms of energy is developed, depends upon what kind of a structure the antecedent energy has acted upon. If radiant energy, so-called, falls upon a mass of matter, what is absorbed is at once transformed into heat or into electric or magnetic effects; *which* one of these depends upon the character of the mechanism upon which the radiant energy acts, but the radiant energy itself, which consists of

ether-waves, is traceable back in every case to
a mass of matter having definite characteristic
motions.

One may therefore say with certainty that every
physical phenomenon is a change in the direction,
or velocity, or character, of the energy present, and
such change has been produced by matter acting as
a transformer.

THE ETHER IS A NON-TRANSFORMER.

It has already been said that the absence of
friction in the ether enables light-waves to maintain
their identity for an indefinite time, and to an
indefinitely great distance. In a uniform, homogeneous substance of any kind, any kind of energy
which might be in it would continue in it without
any change. Uniformity and homogeneity imply
similarity throughout, and the necessary condition
for transformation is unlikeness. One might not
look for any kind of physical phenomenon which
was not due to the presence and activity of some
heterogeneity.

As a ray of light continues a ray of light so long
as it exists in free ether, so all kinds of radiations,
of whatever wave-length, continue identical until
they fall upon some mechanical structure called
matter. Translatory motion continues translatory,
rotary continues rotary, and vibratory continues

to be vibratory, and no transforming change can take place in the absence of matter. The ether is helpless.

13. MATTER IS ELASTIC.

It is commonly stated that certain substances, like putty and dough, are inelastic, while some other substances, like glass, steel, and wood, are elastic. This quality of elasticity, as manifested in such different degrees, depends upon molecular combinations; some of which, as in glass and steel, are favourable for exhibiting it, while others mask it, for the ultimate atoms of all kinds are certainly highly elastic.

The measure of elasticity in a mass of matter is the velocity with which a wave-motion will be transmitted through it. Thus the elasticity of the air determines the velocity of sound in it. If the air be heated, the elasticity is increased and the sound moves faster. The rates of such sound-conduction range from a few feet in a second to about 16,000, five times swifter than a cannon ball. In such elastic bodies as vibrate to and fro like the prongs of a tuning-fork, or give sounds of a definite pitch, the rate of vibration is determined by the size and shape of the body as well as by their elementary composition. The smaller a body is, the higher its vibratory rate, if it be made of the same material and the form remains the same. Thus a tuning-fork, that may be carried in the waistcoat-pocket, may vibrate 500 times a second. If it were only the fifty-millionth of an inch in size, but of the same material and form, it would vibrate 30,000,000000 times a second ; and if it were made of ether, instead of steel, it would vibrate as many times faster as the velocity of waves in the ether is greater than it is in steel, and would be as many as 400,000000,000000 times per second. The amount of displacement, or the amplitude of vibration, with the pocket-fork might be no more than the hundredth of an inch, and this rate measured as translation velocity would be but five inches per second. If the fork were of atomic magnitude, and should swing its sides one half the diameter of the atom, or say the hundred-millionth of an inch, the translational velocity would be equivalent to about eighty miles a second, or a hundred and fifty times the velocity of a cannon ball, which may be reckoned at about 3000 feet.

That atoms really vibrate at the above rate per second is very certain, for their vibrations produce ether-waves the length of which may be accurately measured. When a tuning-fork vibrates 500 times a second, and the sound travels 1100 feet in the same interval, the length of each wave will be found by dividing the velocity in the air by the number of vibrations, or $1100 \div 500 = 2.2$ feet. In like manner, when one knows the velocity and wave-length, he may compute the number of vibrations by dividing the velocity by the wave-length. Now the velocity of the waves called light is 186,000 miles a second, and a light-wave may be one forty thousandth of an inch long. The atom that produces the wave must be vibrating as many times per second as the fifth thousandth of an inch is contained in 186,000 miles. Reducing this number to inches we have

$$\frac{186,000 \times 5280 \times 12}{\frac{1}{40,000}} = 400,000,000,000,000, \text{ nearly.}$$

This shows that the atoms are minute elastic bodies that change their form rapidly when struck. As rapid as the change is, yet the rate of movement is only one-fifth that of a comet when near the sun, and is therefore easily comparable with other velocities observed in masses of matter.

These vibratory motions, due to the elasticity of the atoms, is what constitutes heat.

THE ETHER IS ELASTIC. –

The elasticity of a mass of matter is its ability to recover its original form after that form has been distorted. There is implied that a stress changes its shape and dimensions, which in turn implies a limited mass and relative change of position of

parts and some degree of discontinuity. From
what has been said of the ether as being unlimited,
continuous, and not made of atoms or molecules, it
will be seen how difficult, if not impossible, it is to
conceive how such a property as elasticity, as
manifested in matter, can be attributed to the
ether, which is incapable of deformation, either in
structure or form, the latter being infinitely extended
in every direction and therefore formless. Nevertheless, certain forms of motion, such as light-waves,
move in it with definite velocity, quite independent
of how they originate. This velocity of 186,000 miles
a second so much exceeds any movement of a mass
of matter that the motions can hardly be compared.
Thus if 400 miles per second be the swiftest speed
of any mass of matter known—that of a comet
near the sun—the ether-wave moves 186,000 ÷ 400
= 465 times faster than such comet, and 900,000
times faster than sound travels in air. It is clear
that if this rate of motion depends upon elasticity,
the elasticity must be of an entirely different type
from that belonging to matter, and cannot be defined in any such terms as are employed for matter.

If one considers gravitative phenomena, the difficulty is enormously increased. The orbit of a
planet is never an exact ellipse, on account of the
perturbations produced by the planetary attractions
—perturbations which depend upon the direction

and distance of the attracting bodies. These, however, are so well known that slight deviations are
easily noticed. If gravitative attraction took any
such appreciable time to go from one astronomical
body to another as does light, it would make very
considerable differences in the paths of the planets
and the earth. Indeed, if the velocity of gravitation were less than a million times greater than
that of light, its effects would have been discovered long ago. It is therefore considered that
the velocity of gravitation cannot be less than
186000,000000 miles per second. How much
greater it may be no one can guess. Seeing
that gravitation is ether-pressure, it does not
seem probable that its velocity can be infinite.
However that may be, the ability of the ether to
transmit pressure and various disturbances, evidently depends upon properties so different from
those that enable matter to transmit disturbances
that they deserve to be called by different names.
To speak of the elasticity of the ether may serve to
express the fact that energy may be transmitted at
a finite rate in it, but it can only mislead one's
thinking if he imagines the process to be similar to
energy transmission in a mass of matter. The two
processes are incomparable. No other word has
been suggested, and perhaps it is not needful for
most scientific purposes that another should be

adopted, but the inappropriateness of the one word for the different phenomena has long been felt.

14. MATTER HAS DENSITY.

This quality is exhibited in two ways in matter. In the first, the different elements in their atomic form have different masses or atomic weights. An atom of oxygen weighs sixteen times as much as an atom of hydrogen ; that is, it has sixteen times as much matter, as determined by weight, as the hydrogen atom has, or it takes sixteen times as many hydrogen atoms to make a pound as it takes of oxygen atoms. This is generally expressed by saying that oxygen has sixteen times the density of hydrogen. In like manner, iron has fifty-six times the density, and gold one hundred and ninety-six. The difference is one in the structure of the atomic elements. If one imagines them to be vortex-rings, they may differ in size, thickness, and rate of rotation ; either of these might make all the observed difference between the elements, including their density. In the second way, density implies compactness of molecules. Thus if a cubic foot of air be compressed until it occupies but half a cubic foot, each cubic inch will have twice as many molecules in it as at first. The amount of air per unit volume will have been doubled, the weight will have been doubled, the amount of matter as determined by its weight will have been doubled, and consequently we say its density has been doubled.

If a bullet or a piece of iron be hammered, the molecules are compacted closer together, and a greater number can be got into a cubic inch when so condensed. In this sense, then, density means the number of molecules in a unit of space, a cubic inch or cubic centimeter. There is implied in this latter case that the molecules do not occupy all the available space, that they may have varying degrees of closeness; in other words, matter is discontinuous, and therefore there may be degrees in density.

THE ETHER HAS DENSITY.

It is common to have the degree of density of the ether spoken of in the same way, and for the same reason, that its elasticity is spoken of. The rate of transmission of a physical disturbance, as of a pressure or a wave-motion in matter, is conditioned by its degree of density; that is, the amount of matter per cubic inch as determined by its weight; the greater the density the slower the rate. So if rate of speed and elasticity be known, the density may be computed. In this way the density of the ether has been deduced by noting the velocity of light. The enormous velocity is supposed to prove that its density is very

small, even when compared with hydrogen. This
is stated to be about equal to that of the air at the
height of two hundred and ten miles above the
surface of the earth, where the air molecules are so
few that a molecule might travel for 60,000,000
miles without coming in collision with another
molecule. In air of ordinary density, a molecule
can on the average move no further than about
the two-hundred-and-fifty-thousandth of an inch
without such collision. It is plain the density of
the ether is so far removed from the density of
anything we can measure, that it is hardly
comparable with such things. If, in addition, one
recalls the fact that the ether is homogeneous, that
is all of one kind, and also that it is not composed
of atoms and molecules, then degree of compactness and number of particles per cubic inch have
no meaning, and the term density, if used, can
have no such meaning as it has when applied to
matter. There is no physical conception gained
from the study of matter that can be useful in
thinking of it. As with elasticity, so density is
inappropriately applied to the ether, but there is
no substitute yet offered.

15. MATTER IS HEATABLE.

So long as heat was thought to be some kind of
an imponderable thing, which might retain its

identity whether it were in or out of matter, its real nature was obscured by the name given to it. An imponderable was a mysterious something like a spirit, which was the cause of certain phenomena in matter. Heat, light, electricity, magnetism, gravitation, were due to such various agencies, and no one concerned himself with the nature of one or the other. Bacon thought that heat was a brisk agitation of the particles of substances, and Count Rumford and Sir Humphrey Davy thought they proved that it could be nothing else, but they convinced nobody. Mayer in Germany and Joule in England showed that quantitative relations existed between work done and heat developed, but not until the publication of the book called *Heat as a Mode of Motion,* was there a change of opinion and terminology as to the nature of heat. For twenty years after that it was common to hear the expressions heat, and radiant heat, to distinguish between phenomena in matter and what is now called radiant energy radiations, or simply etherwaves. Not until the necessity arose for distinguishing between different forms of energy, and the conditions for developing them, did it become clear to all that a change in the form of energy implied a change in the form of motion that embodied it. The energy called heat energy was proved to be a vibratory motion of molecules, and what happened in the ether as a result of such vibrations is no longer spoken of as heat, but as ether waves. When it is remembered that the ultimate atoms are elastic bodies, and that they will, if free, vibrate in a periodic manner when struck or shaken in any way, just as a ball will vibrate after it is struck, it is easy to keep in mind the distinction between the mechanical form of motion spent in striking and the vibratory form of the motion produced by it. The latter is called heat; no other form of motion than that is properly called heat.   It is this alone that represents temperature, the rate and amplitude of such atomic and molecular vibrations as constitute change of form.   Where molecules like those in a gas have some freedom of movement between impacts, they bound away from each other with varying velocities.   The path of such motion may be long or short, depending upon the density or compactness of the molecules, but such changes in position are not heat for a molecule any more than the flight of a musket ball is heat, though it may be transformed into heat on striking the target.

This conception of heat as the rapid change in the form of atoms and molecules, due to their elasticity, is a phenomenon peculiar to matter.   It implies a body possessing form that may be changed; elasticity, that its changes may be periodic, and degrees of freedom that secure space for the changes. Such a body may be heated. Its temperature will depend upon the amplitude of such vibrations, and will be limited by the maximum amplitude.

THE ETHER IS UNHEATABLE.

The translatory motion of a mass of matter, big or little, through the ether, is not arrested in any degree so far as observed, but the internal vibratory motion sets up waves in the ether, the ether absorbs the energy, and the amplitude is continually lessened. The, motion has been transferred and transformed ; transferred from matter to the ether, and transformed from vibratory to waves travelling at the rate of 186,000 miles per second. The latter is not heat, but the result of heat. With the ether constituted as described, such vibratory motion as constitutes heat is impossible to it, and hence the characteristic of heat-motion in it is impossible ; it cannot therefore be heated. The space between the earth and the sun may have any assignable amount of energy in the form of ether waves or light, but not any temperature. One might loosely say that the temperature of empty spaces was absolute zero, but that would not be quite correct, for the idea of temperature cannot properly be entertained as applicable to the ether.

To say that its temperature was absolute zero, would serve to imply that it might be higher, which is inadmissible.

When energy has been transformed, the old name by which the energy was called must be dropped. Ether cannot be heated.

16. MATTER IS INDESTRUCTIBLE.

This is commonly said to be one of the essential properties of matter. All that is meant by it, however, is simply this: In no physical or chemical process to which it has been experimentally subjected has there been any apparent loss. The matter experimented upon may change from a solid or liquid to a gas, or the molecular change called chemical may result in new compounds, but the weight of the material and its atomic constituents have not appreciably changed. That matter cannot be annihilated is only the converse of the proposition that matter cannot be created, which ought always to be modified by adding, by physical or chemical processes at present known. A chemist may work with a few grains of a substance in a beaker, or test-tube, or crucible, and after several solutions, precipitations, fusions and dryings, may find by final weighing that he has not lost any appreciable amount, but how much is an appreciable amount? A fragment of matter the ten-

thousandth of an inch in diameter has too small a
weight to be noted in any balance, yet it would be
made up of thousands of millions of atoms. Hence
if, in the processes to which the substance had been
subjected, there had been the total annihilation of
thousands of millions of atoms, such phenomenon
would not have been discovered by weighing.
Neither would it have been discovered if there had
been a similar creation or development of new
matter. All that can be asserted concerning such
events is, that they have not been discovered with
our means of observation.

The alchemists sought to transform one element
into another, as lead into gold. They did not
succeed. It was at length thought to be impossible,
and the attempt to do it an absurdity. Lately,
however, telescopic observation of what is going on
in nebula, which has already been referred to, has
somewhat modified ideas of what is possible and
impossible in that direction. It is certainly
possible roughly to conceive how such a structure
as a vortex-ring in the ether might be formed.
With certain polarizing apparatus it is possible to
produce rays of circularly polarized light. These
are rays in which the motion is an advancing rotation like the wire in a spiral spring. If such a line
of rotations in the ether were flexible, and the two
ends should come together, there is reason for

E

thinking they would weld together, in which case the structure would become a vortex-ring and be as durable as any other. There is reason for believing, also, that somewhat similar movements are always present in a magnetic field, and though we do not know how to make them close up in the proper way, it does not follow that it is impossible for them to do so.

The bearing of all this upon the problem of the transmutation of elements is evident. No one now will venture to deny its possibility as strongly as it was denied a generation ago. It will also lead one to be less confident in the theory that matter is indestructible. Assuming the vortex-ring theory of atoms to be true, if in any way such a ring could be cut or broken, there would not remain two or more fragments of a ring or atom. The whole would at once be dissolved into the ether. The ring and rotary energy that made it an atom would be destroyed, but not the substance it was made of, nor the energy which was embodied therein. For a long time philosophers have argued, and commonsense has agreed with them, that an atom which could not be ideally broken into two parts was impossible, that one could at any rate think of half an atom as a real objective possibility. This vortex-ring theory shows easily how possible it is to-day to think what once was philosophically incredible. It shows that metaphysical reasoning may be ever so clear and apparently irrefragable, yet for all that it may be very unsound. The trouble does not come so much from the logic as from the assumption upon which the logic is founded. In this particular case the assumption was that the ultimate particles of matter were hard, irrefragable somethings, without necessary relations to anything else, or to energy, and irrefragable only because no means had been found of breaking them.

The destructibility or indestructibility of the ether cannot be considered from the same standpoint as that for matter, either ideally or really. Not ideally, because we are utterly without any mechanical conceptions of the substance upon which one can base either reason or analogy; and not really, because we have no experimental evidence as to its nature or mode of operation. If it be continuous, there are no interspaces, and if it be illimitable there is no unfilled space anywhere. Furthermore, one might infer that if in any way a portion of the ether could be annihilated, what was left would at once fill up the vacated space, so there would be no record left of what had happened. Apparently, its destruction would be the destruction of a substance, which is a very different thing from the destruction of a mode of motion. In the latter, only the form of the motion need be destroyed to

completely obliterate every trace of the atom. In
the former, there would need to be the destruction
of both substance and energy, for it is certain, for
reasons yet to be attended to, that the ether is
saturated with energy.

One may, without mechanical difficulties, imagine
a vortex-ring destroyed. It is quite different with
the ether itself, for if it were destroyed in the same
sense as the atom of matter, it would be changed
into something else which is not ether, a proposition which assumes the existence of another entity,
the existence for which is needed only as a mechanical antecedent for the other. The same assumption would be needed for this entity as for the
ether, namely, something out of which it was made,
and this process of assuming antecedents would be
interminable. The last one considered would have
the same difficulties to meet as the ether has now.
The assumption that it was in some way and at
some time created is more rational, and therefore
more probable, than that it either created itself or
that it always existed. Considered as the underlying stratum of matter, it is clear that changes
of any kind in matter can in no way affect the
quantity of ether.

17. MATTER HAS INERTIA.

The resistance that a mass of matter opposes to a change in its position or rate and direction of movement, is called inertia. That it should actively oppose anything has been already pointed out as reason for denying that matter is inert, but inertia is the measure of the reaction of a body when it is acted upon by pressure from any source tending to disturb its condition of either rest or motion. It is the equivalent of mass, or the amount of matter as measured by gravity, and is a fixed quantity; for inertia is as inherent as any other quality, and belongs to the ultimate atoms and every combination of them. It implies the ability to absorb energy, for it requires as much energy to bring a moving body to a standstill as was required to give it its forward motion.

Both rotary and vibratory movements are opposed by the same property. A grindstone, a tuning-fork, and an atom of hydrogen require, to move them in their appropriate ways, an amount of energy proportionate to their mass or inertia, which energy is again transformed through friction into heat and radiated away.

One may say that inertia is the measure of the ability of a body to transfer or transform mechanical energy. The meteorite that falls upon the earth to-day gives, on its impact, the same amount of energy it would have given if it had struck the earth ten thousand years ago. The inertia of the meteor has persisted, not as energy, but as a factor of energy. We commonly express the energy of a mass of matter by $\dfrac{mv^2}{2}$, where m stands for the mass and v for its velocity. We might as well, if it were as convenient, substitute inertia for mass, and write the expression $\dfrac{iv^2}{2}$, for the mass, being measured by its inertia, is only the more common and less definitive word for the same thing. The energy of a mass of matter is, then, proportional to its inertia, because inertia is one of its factors. Energy has often been treated as if it were an objective thing, an entity and a unity; but such a conception is evidently wrong, for, as has been said before, it is a product of two factors, either of which may be changed in any degree if the other be changed inversely in the same degree. A cannon ball weighing 1000 pounds, and moving 100 feet per second, will have 156,000 foot-pounds of energy, but a musket ball weighing an ounce will have the same amount when its velocity is 12,600 feet per second. Nevertheless, another body acting upon either bullet or cannon ball, tending to move either in some new direction, will be as efficient while those bodies are moving at any assignable rate as when they are quiescent, for the change in direction will depend upon the inertia of the bodies, and that is constant.

The common theory of an inert body is one that is wholly passive, having no power of itself to move or do anything, except as some agency outside itself compels it to move in one way or another, and thus endows it with energy. Thus a stone or an iron nail are thought to be inert bodies in that sense, and it is true that either of them will remain still in one place for an indefinite time and move from it only when some external agency gives them impulse and direction. Still it is known that such bodies will roll down hill if they will not roll up, and each of them has itself as much to do with the down-hill movement as the earth has; that is, it attracts the earth as much as the earth attracts it. If one could magnify the structure of a body until the molecules became individually visible, every one of them would be seen to be in intense activity, changing its form and relative position an enormous number of times per second in undirected ways. No two such molecules move in the same way at the same time, and as all the molecules cohere together, their motions in different directions balance each other, so that the body as a whole does not change its position, not because there is no moving agency in itself, but because the individual movements are scattering, and not in a common direction. An army may remain in one place for a long time. To one at a distance it is quiescent, inert. To one in the camp there is abundant sign of activity, but the movements are individual movements, some in one direction and some in another, and often changing. The same army on the march has the same energy, the same rate of individual movement; but all have a common direction, it moves as a whole body into new territory. So with the molecules of matter. In large masses they appear to be inert, and to do nothing, and to be capable of doing nothing. That is only due to the fact that their energy is undirected, not that they can do nothing. The inference that if quiescent bodies do not act in particular ways they are inert, and cannot act in any kind of a way, is a wrong inference. An illustration may perhaps make this point plainer. A lump of coal will be still as long as anything if it be undisturbed. Indeed, it has thus lain in a coalbed for millions of years probably, but if coal be placed where it can combine with oxygen, it forthwith does so, and during the process yields a large amount of energy in the shape of heat. One pound of coal in this way gives out 14,000 heat units, which is the equivalent of 11,000,000 foot-pounds of work, and if it could be all utilized would furnish a horse-power for five and a half hours. Can any inert body weighing a pound furnish a horse-power for half a day? And can a body give out what it has not got? Are gunpowder and nitro-glycerine inert? Are bread and butter and foods in general inert because they will not push and pull as a man or a horse may? All have energy, which is available in certain ways and not in others, and whatever possesses energy available in any way is not an ideally inert body. Lastly, how many inert bodies together will it take to make an active body? If the question be absurd, then all the phenomena witnessed in bodies, large or small, are due to the fact that the atoms are not inert, but are immensely energetic, and their inertia is the measure of their rates of exchanging energy.

THE ETHER IS CONDITIONALLY POSSESSED OF INERTIA.

A moving mass of matter is brought to rest by friction, because it imparts its motion at some rate to the body it is in contact with. Generally the energy is transformed into heat, but sometimes it appears as electrification. Friction is only possible because one or both of the bodies possess inertia. That a body may move in the ether for an indefinite time without losing its velocity has been stated as a reason for believing the ether to be frictionless. If it be frictionless, then it is without inertia, else the energy of the earth and of a ray of light would be frittered away. A ray of light can only be transformed when it falls upon molecules which may be heated by it. As the ether cannot be heated and cannot transform translational energy, it is without inertia for *such* a form of motion and its embodied energy.

It is not thus with other forms of energy than the translational. Atomic and molecular vibrations are so related to the ether that they are transformed into waves, which are conducted away at a definite rate. This shows that such property of inertia as is possessed by the ether is selective and not like that of matter, which is equally "inertiative" under all conditions. Similarly with electric and magnetic phenomena, it is capable of transforming the energy which may reside as stress in the ether, and other bodies moving in the space so affected meet with frictional resistance, for they become heated if the motion be maintained. On the other hand, there is no evidence that the body which produced the electric or magnetic stress suffers any degree of friction on moving in precisely the same space. A bar magnet rotating on its longitudinal axis does not disturb its own field, but a piece

of iron revolving near the magnet will not only
become heated, but will heat the stationary magnet.
Much experimental work has been done to discover, if possible, the relation of a magnet to its
ether field. As the latter is not disturbed by the
rotation of the magnet, it has been concluded
that the field does not rotate ; but as every
molecule in the magnet has its own field independent of all the rest, it is mechanically probable
that each such field does vary in the rotation,
but among the thousands of millions of such fields
the average strength of the field does not vary within
measurable limits. Another consideration is that
the magnetic field itself, when moved in space,
suffers no frictional resistance. There is no magnetic energy wasted through ether inertia. These
phenomena show that whether the ether exhibits
the quality called inertia depends upon the kind
of motion it has.

18. MATTER IS MAGNETIC.

The ordinary phenomenon of magnetism is
shown by bringing a piece of iron into the neighbourhood of a so-called magnet, where it is attracted
by the latter, and if free to move will go to and
cling to the magnet. A delicately suspended
magnetic needle will be affected appreciably by a
strong magnet at the distance of several hundred

feet. As the strength of such action varies inversely as the square of the distance from the magnet, it is evident there can be no absolute boundary to it. At a distance from an ordinary magnet it becomes too weak to be detected by our methods, not that there is a limit to it. It is customary to think of iron as being peculiarly endowed with magnetic quality, but all kinds of matter possess it in some degree. Wood, stone, paper, oats, sulphur, and all the rest, are attracted by a magnet, and will stick to it if the magnet be a strong one. Whether a piece of iron itself exhibits the property depends upon its temperature, for near 700 degrees it becomes as magnetically indifferent as a piece of copper at ordinary temperature. Oxygen, too, at 200 degrees below the zero of Centigrade adheres to a magnet like iron.

In this as in so many other particulars, how a piece of matter behaves depends upon its temperature, not that the essential qualities are modified in any degree, but temperature interferes with atomic arrangement and aggregation, and so disguises their phenomena.

As every kind of matter is thus affected by a magnet, the manifestations differing but in degree, it follows that all kinds of atoms—all the elements —are magnetic. An inherent property in them, as much so as gravitation or inertia ; apparently a quality depending upon the structure of the atoms themselves, in the same sense as gravitation is thus dependent, as it is not a quality of the ether.

An atom must, then, be thought of as having polarity, different qualities on the two sides, and possessing a magnetic field as extensive as space itself. The magnetic field is the stress or pressure in the ether produced by the magnetic body. This ether pressure produced by a magnet may be as great as a ton per square inch. It is this pressure that holds an armature to the magnet. As heat is a molecular condition of vibration, and radiant energy the result of it, so is magnetism a property of molecules, and the magnetic field the temporary condition in the ether, which depends upon the presence of a magnetic body. We no longer speak of the wave-motion in the ether which results from heat, as heat, but call it radiation, or ether waves, and for a like reason the magnetic field ought not to be called magnetism.

THE ETHER IS NON-MAGNETIC.

A magnetic field manifests itself in a way that implies that the ether structure, if it may be said to have any, is deformed—deformed in such a sense that another magnet in it tends to set itself in the plane of the stress; that is, the magnet is twisted into a new position to accommodate itself to the condition of the medium about it. The new position is the result of the reaction of the ether upon the magnet and ether pressure acting at right angles to the body that produced the stress. Such an action is so anomalous as to suggest the propriety of modifying the so-called third law of motion, viz., action and reaction are equal and opposite, adding that sometimes action and reaction are at right angles.

There is no condition or property exhibited by the ether itself which shows it to have any such characteristic as attraction, repulsion, or differences in stress, except where its condition is modified by the activities of matter in some way. The ether itself is not attracted or repelled by a magnet; that is, it is not a magnetic body in any such sense as matter in any of its forms is, and therefore cannot properly be called magnetic.

It has been a mechanical puzzle to understand how the vibratory motions called heat could set up light waves in the ether seeing that there is an absence of friction in the latter. In the endeavour to conceive it, the origin of sound-waves has been in mind, where longitudinal air-waves are produced by the vibrations of a sounding body, and molecular impact is the antecedent of the waves. The analogy does not apply. The following exposition may be helpful in grasping the idea of such transformation and change of energy from matter to the ether.

Consider a straight bar permanent magnet to be held in the hand. It has its north and south poles and its field, the latter extending in every direction to an indefinite distance. The field is to be considered as ether stress of such a sort as to tend to set other magnets in it in new positions. If at a distance of ten feet there were a delicately-poised magnet needle, every change in the position of the magnet held in the hand would bring about a change in the position of the needle. If the position of the hand magnet were completely reversed, so the south pole faced where the north pole faced before, the field would have been completely reversed, and the poised needle would have been pushed by the field into an opposite position. If the needle were a hundred feet away, the change would have been the same except in amount. The same might be said if the two were a mile apart, or the distance of the moon or any other distance, for there is no limit to an ether magnetic field. Suppose the hand magnet to have its direction completely reversed once in a second. The whole field, and the direction of the stress, would necessarily be reversed as often. But this kind of change in stress is known by experiment to travel with the speed of light, 186,000 miles a second ; the disturbance due to the change of position of the magnet will therefore be felt in some degree throughout space. In a second and a third of a second it will have reached the moon, and a magnet there will be in some measure affected by it. If there were an observer there with a delicate-enough magnet, he could be witness to its changes once a second for the same reason one in the room could. The only difference would be one of amount of swing. It is therefore theoretically possible to signal to the moon with a swinging magnet. Suppose again that the magnet should be swung twice a second, there would be formed two waves, each one half as long as the first. If it should swing ten times a second, then the waves would be one-tenth of 186,000 miles long. If in some mechanical way it could be rotated 186,000 times a second, the wave would be but one mile long. Artificial ways have been invented for changing this magnet field as many as 100 million times a second, and the corresponding wave is less than a foot long. The shape of a magnet does not necessarily make it weaker or stronger as a magnet, but if the poles are near together the magnetic field is denser between them than when they are separated. The ether stress is differently distributed for every change in the relative positions of the poles.

A common U-magnet, if struck, will vibrate like a tuning-fork, and gives out a definite pitch. Its poles swing towards and away from each other at uniform rates, and the pitch of the magnet will depend upon its size, thickness, and the material it is made of.

Let ten or fifteen ohms of any convenient-sized wire be wound upon the bend of a commercial U-magnet. Let this wire be connected to a telephone in its circuit. When the magnet is made to sound like a tuning-fork, the pitch will be reproduced in the telephone very loudly. If another magnet with a different pitch be allowed to vibrate near the former, the pitch of the vibrating body will be heard in the telephone, and these show that the changing.magnetic field reacts upon the quiescent magnet, and compels the latter to vibrate at the same rate. The action is an ether action, the waves are ether waves, but they are relatively very long. If the magnet makes 500 vibrations a second, the waves will be 372 miles long, the number of times 500 is contained in 186,000 miles. Imagine the magnet to become smaller and smaller until it was the size of an atom, the one-fifty-millionth of an inch. Its vibratory rate would be proportionally increased, and changes in its form will still bring about changes in its magnetic field. But its magnetic field is practically limitless, and the number of vibrations per second is to be reckoned as millions of millions; the waves are

F

correspondingly short, small fractions of an inch. When they are as short as the one-thirty-seven-thousandth of an inch, they are capable of affecting the retina of the eye, and then are said to be visible as red light. If the vibratory rate be still higher, and the corresponding waves be no more than one-sixty-thousandth of an inch long, they affect the retina as violet light, and between these limits there are all the waves that produce a complete spectrum. The atoms, then, shake the ether in this way because they all have a magnetic hold upon the ether, so that any disturbance of their own magnetism, such as necessarily comes when they collide, reacts upon the ether for the same reason that a large magnet acts thus upon it when its poles approach and recede from each other. It is not a phenomenon of mechanical impact or frictional resistance, since neither are possible in the ether.

19. MATTER EXISTS IN SEVERAL STATES.

Molecular cohesion exists between very wide ranges. When strong, so if one part of a body is moved the whole is moved in the same way, without breaking continuity or the relative positions of the molecules, we call the body a solid. In a liquid, cohesion is greatly reduced, and any part of it may be deformed without materially changing the form of the rest. The molecules are free to move about each other, and there is no definite position which any need assume or keep. With gases, the molecules are without any cohesion, each one is independent of every other one, collides with and bounds away from others as free elastic particles do. Between impacts it moves in what is called its free path, which may be long or short as the density of the gas be less or greater.

These differing degrees of cohesion depend upon temperature, for if the densest and hardest substances are sufficiently heated they will become gaseous. This is only another way of saying that the states of matter depend upon the amount of molecular energy present. Solid ice becomes water by the application of heat. More heat reduces it to steam; still more decomposes the steam molecules into oxygen and hydrogen molecules; and lastly, still more heat will decompose these molecules into their atomic state, complete dissociation. On cooling, the process of reduction will be reversed until ice has been formed again.

Cohesive strength in solids is increased by reduction of temperature, and metallic rods become stronger the colder they are.

No distinction is now made between cohesion and chemical affinity, and yet at low temperatures chemical action will not take place, which phenomenon shows there is a distinction between molecular cohesion and molecular structure. In molecular structure, as determined by chemical activity, the molecules and atoms are arranged in definite ways which depend upon the rate of vibrations of the components. The atoms are set in definite positions to constitute a given molecule. But atoms or molecules may cohere for other reasons, gravitative or magnetic, and relative positions would be immaterial. In the absence of temperature, a solid body would be solider and stronger than ever, while a gaseous mass would probably fall by gravity to the floor of the containing vessel like so much dust. The molecular structure might not be changed, for there would be no agency to act upon it in a disturbing way.

THE ETHER HAS NO CORRESPONDING STATES.

Degrees of density have already been excluded, and the homogeneity and continuity of the ether would also exclude the possibility of different states at all comparable with such as belong to matter. As for cohesion, it is doubtful if the term ought to be applied to such a substance. The word itself seems to imply possible separateness, and if the ether be a single indivisible substance, its cohesion must be infinite and is therefore not a matter of degree. The ether has sometimes been considered as an elastic solid, but such solidity is comparable with nothing we call solid in matter, and the word has to be defined in a special sense in order that its use may be tolerated at all. In addition to this, some of the phenomena exhibited by it, such as diffraction and double refraction, are quite incompatible with the theory that the ether is an elastic solid. The reasons why it cannot be considered as a liquid or gas have been considered previously.

The expression *states of matter* cannot be applied to the ether in any such sense as it is applied to matter, but there is one sense when possibly it may be considered applicable. Let it be granted that an atom is a vortex-ring of ether in the ether, then the state of being in ring rotation would suffice to differentiate that part of the ether from the rest, and give to it a degree of individuality not possessed by the rest ; and such an atom might be called a state of ether. In like manner, if other forms of motion, such as transverse waves, circular and elliptical spirals, or others, exist in the ether, then such movements give special character to the part thus active, and it would be proper to speak of such states of the ether, but even thus the word would not be used in the same sense as it is used when one speaks of the states of matter as being solid, liquid, and gaseous.

20. SOLID MATTER CAN EXPERIENCE A SHEARING STRESS, LIQUIDS AND GASES CANNOT.

A sliding stress applied to a solid deforms it to a degree which depends upon the stress and the degree of rigidity preserved by the body. Thus if the hand be placed upon a closed book lying on the table, and pressure be so applied as to move the upper side of the book but not the lower, the book is said to be subject to a shearing stress. If the pressing hand has a twisting motion, the book will be warped. Any solid may be thus sheared or warped, but neither liquids nor gases can be so affected. Molecular cohesion makes it possible in the one, and the lack of it, impossible in the others. The solid can maintain such a deformation indefinitely long, if the pressure does not rupture its molecular structure.

THE ETHER CAN MAINTAIN A SHEARING STRESS.

The phenomena in a magnetic field show that the stress is of such a sort as to twist into a new directional position the body upon which it acts as exhibited by a magnetic needle, also as indicated by the transverse vibrations of the ether waves, and again by the twist given to plane polarized light when moving through a magnetic field. These are all interpreted as indicative of the direction of ether stress, as being similar to a shearing stress in solid matter. The fact has been adduced to show the ether to be a solid, but such a phenomenon is certainly incompatible with a liquid or gaseous ether. This kind of stress is maintained indefinitely about a permanent magnet, and the mechanical pressure which may result from it is a measure of the strength of the magnetic field, and may exceed a thousand pounds per square inch.

21. OTHER PROPERTIES OF MATTER.

There are many secondary qualities exhibited by matter in some of its forms, such as hardness, brittleness, malleability, colour, etc., and the same ultimate element may exhibit itself in the most diverse ways, as is the case with carbon, which exists as lamp-black, charcoal, graphite, jet, anthracite and diamond, ranging from the softest to the hardest of known bodies. Then it may be black or colourless. Gold is yellow, copper red, silver white, chlorine green, iodine purple. The only significance any or all of such qualities have for us here is that the ether exhibits none of them. There is neither hardness nor brittleness, nor colour, nor any approach to any of the characteristics for the identification of elementary matter.

22. SENSATION DEPENDS UPON MATTER.

However great the mystery of the relation of body to mind, it is quite true that the nervous system is the mechanism by and through which all sensation comes, and that in our experience in the absence of nerves there is neither sensation nor consciousness. The nerves themselves are but complex chemical structures ; their molecular constitution is said to embrace as many as 20,000 atoms, chiefly carbon, hydrogen, oxygen, and nitrogen. There must be continuity of this structure too, for to sever a nerve is to paralyze all beyond. If all knowledge comes through experience, and all experience comes through the nervous system, the possibilities depend upon the mechanism each one is provided with for absorbing from his environment, what energies there are that can act upon the nerves. Touch, taste, and smell imply contact, sound has greater range, and sight has the immensity of the universe for its field. The most distant but visible star acts through the optic nerve to present itself to consciousness. It is not the ego that looks out through the eyes, but it is the universe that pours in upon the ego.

Again, all the known agencies that act upon the nerves, whether for touch or sound or sight, imply matter in some of its forms and activities, to adapt the energy to the nervous system. The mechanism for the perception of light is complicated. The light acts upon a sensitive surface where molecular structure is broken up, and this disturbance is in the presence of nerve terminals, and the sensation is not in the eye but in the sensorium. In like manner for all the rest ; so one may fairly say that matter is the condition for sensation, and in its absence there would be nothing we call sensation.

THE ETHER IS INSENSIBLE TO NERVES.

The ether is in great contrast with matter in this particular. There is no evidence that in any direct way it acts upon any part of the nervous system, or upon the mind. It is probable that this lack of relation between the ether and the nervous system was the chief reason why its discovery was so long delayed, as the mechanical necessities for it even now are felt only by such as recognize continuity as a condition for the transmission of energy of whatever kind it may be. Action at a distance contradicts all experience, is philosophically incredible, and is repudiated by every one who once perceives that energy has two factors—substance and motion.

The table given below presents a list of twenty-two of the known properties of matter contrasted with those exhibited by the ether. In none of them are the properties of the two identical, and in most of them what is true for one is not true for the other. They are not simply different, they are incomparable.

From the necessities of the case, as knowledge has been acquired and terminology became essential for making distinctions, the ether has been described in terms applicable to matter, hence such terms as mass, solidity, elasticity, density, rigidity, etc., which have a definite meaning and convey definite mechanical conceptions when applied to matter, but have no corresponding meaning and convey no such mechanical conceptions when applied to the ether. It is certain that they are inappropriate, and that the ether and its properties cannot be described in terms applicable to matter. Mathematical considerations derived from the study of matter have no advantage, and are not likely to lead us to a knowledge of the ether.

Only a few have perceived the inconsistency of thinking of the two in the same terms. In his *Grammar of Science*, Prof. Karl Pearson says, " We find that our sense-impressions of hardness, weight, colour, temperature, cohesion, and chemical constitution, may all be described by the aid of the motions of a single medium, which itself is conceived to have no hardness, weight, colour, temperature, nor indeed elasticity of the ordinary conceptua type."

None of the properties of the ether are such as

one would or could have predicted if he had had all the knowledge possessed by mankind. Every phenomenon in it is a surprise to us, because it does not follow the laws which experience has enabled us to formulate for matter. A substance which has none of the phenomenal properties of matter, and is not subject to the known laws of matter, ought not to be called matter. Ether phenomena and matter phenomena belong to different categories, and the ends of science will not be conserved by confusing them, as is done when the same terminology is employed for both.

There are other properties belonging to the ether more wonderful, if possible, than those already mentioned. Its ability to maintain enormous stresses of various kinds without the slightest evidence of interference. There is the gravitational stress, a direct pull between two masses of matter. Between two molecules it is immeasurably small even when close together, but the prodigious number of them in a bullet brings the action into the field of observation, while between such bodies as the earth and moon or sun, the quantity reaches an astonishing figure. Thus if the gravitative tension due to the gravitative attraction of the earth and moon were to be replaced by steel wires connecting the two bodies to prevent the moon from leaving its orbit, there would be needed four

number ten steel wires to every square inch upon
the earth, and these would be strained nearly to the
breaking point.   Yet this stress is not only endured
continually by this pliant, impalpable, transparent
medium, but other bodies can move through the
same space apparently as freely as if it were entirely
free.   In addition to this, the stress from the sun
and the more variable stresses from the planets are
all endured by the same medium in the same space
and apparently a thousand or a million times more
would not make the slightest difference.   Rupture
is impossible.

Electric and magnetic stresses, acting parallel or
at right angles to the other, exist in the same space
and to indefinite degrees, neither modifying the
direction nor amount of either of the others.

These various stresses have been computed to
represent energy, which if it could be utilized, each
cubic inch of space would yield five hundred horse-
power.   It shows what a store-house of energy the
ether is.   If every particle of matter were to be
instantly annihilated, the universe of ether would
still have an inexpressible amount of energy left.
To draw at will directly from this inexhaustible
supply, and utilize it for the needs of mankind, is not
a forlorn hope.

The accompanying table presents these con-
trasting properties for convenient inspection,

CONTRASTED PROPERTIES OF MATTER AND THE ETHER.

MATTER.	ETHER.
1. Discontinuous...............	Continuous
2. Limited	Unlimited
3. Heterogeneous	Homogeneous
4. Atomic........................	Non-atomic
5. Definite structure	Structureless
6. Gravitative	Gravitationless
7. Frictionable.................	Frictionless
8. Aeolotropic	Isotropic
9. Chemically selective	————
10. Harmonically related ...	————
11. Energy embodied	Energy endowed
12. Energy transformer	Non-transformer
13. Elastic	Elastic?
14. Density.......................	Density?
15. Heatable	Unheatable
16. Indestructible?	Indestructible
17. Inertiative	Inertiative conditionally
18. Magnetic	————
19. Variable states	————
20. Subject to shearing stress in solid	Shearing stress maintained
21. Has Secondary qualities	————
22. Sensation depends upon	Insensible to nerves

CHAPTER III

So far as we have knowledge to-day, the only factors we have to consider in explaining physical phenomena are : (1) Ordinary matter, such as con-stitutes the substance of the earth, and the heavenly bodies ; (2) the ether, which is omnipresent ; and (3) the various forms of motion, which are mutually transformable in matter, and some of which, but not all, are transformable into ether forms. For in-stance, the translatory motion of a mass of matter can be imparted to another mass by simple impact, but translatory motion cannot be imparted to the ether, and, for that reason, a body moving in it is not subject to friction, and continues to move on with velocity undiminished for an indefinite time ; but the vibratory motion which constitutes heat is transformable into wave motion in the ether, and is transmitted away with the speed of light. The kind of motion which is thus transformed is not even a to-and-fro swing of an atom, or molecule, like the swing of a pendulum bob, but that due to a change of form of the atoms within the molecule, otherwise there could be no such thing as spectrum analysis. Vibratory motion of the matter becomes undulatory motion in the ether. The vibratory motion we call heat ; the wave-motion we call sometimes radiant energy, sometimes light. Neither of these terms is a good one, but we now have no others.

It is conceded that it is not proper to speak of the wave-motion in the ether as *heat ;* it is also admitted that the ether is not heated by the presence of the wave—or, in other words, the temperature of the ether is absolute zero. Matter only can be heated. But the ether waves can heat other matter they may fall on ; so there are three steps in the process and two transformations—(1) vibrating matter; (2) waves in the ether ; (3) vibration in other matter. Energy has been transferred indirectly. What is important to bear in mind is, that when a form of energy in matter is transformed in any manner so as to lose its characteristics, it is not proper to call it by the same name after as before, and this we do in all cases when the transformation is from one kind in matter to another kind in matter. Thus, when a bullet is shot against a target, before it strikes it has what we call mechanical energy, and we measure that in foot-pounds; after it has struck the target, the transformation is into heat, and this has its mechanical equivalent, but is not called mechanical energy, nor are the motions which embody it similar. The mechanical ideas in these phenomena are easy to grasp. They apply to the phenomena of the mechanics of large and small bodies, to sound, to heat, and to light, as ordinarily considered, but they have not been applied to electric phenomena, as they evidently should be, unless it be held that such phenomena are not related to ordinary pheno-mena, as the latter are to one another.

When we would give a complete explanation of the phenomena exhibited by, say, a heated body, we need to inquire as to the antecedents of the manifestation, and also its consequents. Where and how did it get its heat? Where and how did it lose it? When we know every step of those processes, we know all there is to learn about them. Let us undertake the same thing for some electrical phenomena.

First, under what circumstances do electrical phenomena arise?

(1) *Mechanical*, as when two different kinds of matter are subject to friction.

(2) *Thermal*, as when two substances in molecular contact are heated at the junction.

(3) *Magnetic*, as when any conductor is in a changing magnetic field.

(4) *Chemical*, as when a metal is being dissolved in any solution.

FIG. 5.—Frictional electrical machine.

(5) *Physiological*, as when a muscle contracts.

Each of these has several varieties, and changes may be rung on combinations of them, as when mechanical and magnetic conditions interact.

(1) In the first case, ordinary mechanical or translational energy is spent as friction, an amount measurable in foot-pounds, and the factors we

G

know, a pressure into a distance. If the surface be
of the same kind of molecules, the whole energy is
spent as heat, and is presently radiated away. If
the surfaces are of unlike molecules, the product is'
a compound one, part heat, part electrical. What
we have turned into the machine we know to be a

Fig. 6.—Thermo-pile.

particular mode of motion. We have not changed
the amount of matter involved ; indeed, we asume,
without specifying and without controversy, that
matter is itself indestructible, and the product,
whether it be of one kind or another, can only be
some form of motion. Whether we can describe
it or not is immaterial ; but if we agree that heat

is vibratory molecular motion, and there be any other kind of a product than heat, it too must also be some other form of motion. So if one is to form a conception of the mechanical origin of

FIG. 7.—Dynamo.

electricity, this is the only one he can have— transformed motion.

(2) When heat is the antecedent of electricity, as in the thermo-pile, that which is turned into the pile we know to be molecular motion of a definite kind. That which comes out of it must be some equivalent

motion, and if all that went in were transformed, then all that came out would be transformed, call it by what name we will and let its amount be what it may.

(3) When a conductor is moved in a magnetic field, the energy spent is measurable in foot-pounds, as before, a pressure into a distance. The energy appears in a new form, but the quantity of matter being unchanged, the only changeable factor is the kind of motion, and that the motion is molecular is evident, for the molecules are heated. Mechanical or mass motion is the antecedent, molecular heat motion is the consequent, and the way we know there has been some intermediate form is, that heat is not conducted at the rate which is observed in such a case. Call it by what name one will, some form of motion has been immediate between the antecedent and the consequent, else we have some other factor of energy to reckon with than ether, matter and motion.

(4) In a galvanic battery, the source of electricity is chemical action; but what is chemical action? Simply an exchange of the constituents of molecules—a change which involves exchange of energy. Molecules capable of doing chemical work are loaded with energy. The chemical products of battery action are molecules of different constitution, with smaller amounts of energy as

measured in calorics or heat units. If the results of
the chemical reaction be prevented from escaping,
by confining them to the cell itself, the whole energy
appears as heat and raises the temperature of the cell.
If a so-called circuit be provided, the energy is dis-

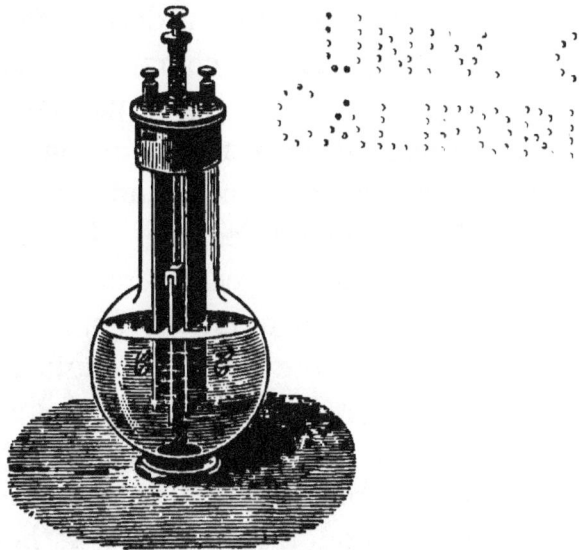

FIG. 8 —Galvanic Battery.

tributed through it, and less heat is spent in the cell,
but whether it be in one place or another, the mass
of matter involved is not changed, and the variable
factor is the motion, the same as in the other cases.
The mechanical conceptions appropriate are the
transformation of one kind of motion into another
kind by the mechanical conditions provided.

(5) Physiological antecedents of electricity are exemplified by the structure and mode of operation of certain muscles (Fig. 9, *a*) in the torpedo and other electrical animals. The mechanical contraction of them results in an electrical excitation, and, if a proper circuit be provided, in an electric current. The energy of a muscle is derived from food, which is itself but a molecular compound loaded with energy of a kind available for muscular transformation. Bread-and-butter has more available energy, pound for pound, than has coal, and can be substituted for coal for running an engine. It is not used, because it costs so much more. There is nothing different, so far as the factors of energy go, between the food of an animal and the food of an engine. What becomes of the energy depends upon the kind of structure it acts on. It may be changed into translatory, and the whole body moves in one direction; or into molecular, and then appears as heat or electrical energy.

If one confines his attention to the only variable factor in the energy in all these cases, and traces out in each just what happens, he will have only motions of one sort or another, at one rate or another, and there is nothing mysterious which enters into the processes.

We will turn now to the mode in which electricity manifests itself, and what it can do. It may

be well to point out at the outset what has occasion-
ally been stated, but which has not received the
philosophical attention it deserves—namely, that

FIG. 9.—Torpedo.

electrical phenomena are reversible; that is, any
kind of a physical process which is capable of
producing electricity, electricity is itself able to
produce. Thus to name a few: If mechanical

motion develops electricity, electricity will produce
mechanical motion ; the movement of a pith ball
and an electric motor are examples.   If chemical
action can produce it, it will produce chemical
action, as in the decomposition of water and electro-
plating.   As heat may be its antecedent, so will it
produce heat.   If magnetism be an antecedent
factor, magnetism may be its product.   What is
called induction may give rise to it in an adjacent

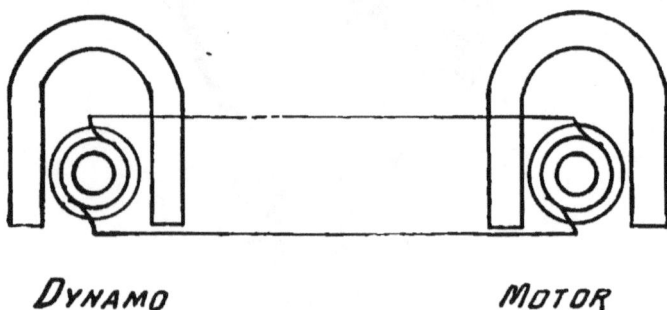

DYNAMO   MOTOR

FIG. 10.—Dynamo and Motor.

conductor, and, likewise, induction may be its
effect.

Let us suppose ourselves to be in a building in
which a steam-engine is at work.   There is fuel,
the furnace, the boiler, the pipes, the engine with
its fly-wheel turning.   The fuel burns in the
furnace, the water is superheated in the boiler, the
steam is directed by the pipes, the piston is moved
by the steam pressure, and the fly-wheel rotates

because of proper mechanism between it and the
piston. No one who has given attention to the
successive steps in the process is so puzzled as to
feel the need of inventing a particular force, or a
new kind of matter, or any agency, at any stage of
the process, different from the simple mechanical
ones represented by a push or a pull. Even if he
cannot see clearly how heat can produce a push, he
does not venture to assume a genii to do the work,
but for the time is content with saying that if he
starts with motion in the furnace and stops with the
motion of the fly-wheel, any assumption of any
other factor than some form of motion between the
two would be gratuitous. He can truthfully say
that he understands the *nature* of that which goes
on between the furnace and the wheel; that it is
some sort of motion, the particular kind of which
he might make out at his leisure.

Suppose once more that, across the road from
an engine-house, there was another building, where
all sorts of machines—lathes, planers, drills, etc.—
were running, but that the source of the power for all
this was out of sight, and that one could see no
connection between this and the engine on the other
side of the street. Would one need to suppose
there was anything mysterious between the two—a
force, a fluid, an immaterial something? This
question is put on the supposition that one should

not be aware of the shaft that might be between the two buildings, and that it was not obvious on simple inspection how the machines got their motions from the engine. No one would be puzzled because he did not know just what the intervening mechanism might be. If the boiler were in the one building, and the engine in the other with the machines, he could see nothing moving between them, even if the steam-pipes were of glass. If matter of any kind were moving, he could not see it there. He would say there *must* be something moving, or pressure could not be transferred from one place to the other.

Substitute for the furnace and boiler a galvanic battery or a dynamo ; for the machines of the shop, one or more motors with suitable wire connections. When the dynamo goes the motors go ; when the dynamo stops the motors stop; nothing can be seen to be turning or moving in any way between them. Is there any necessity for assuming a mysterious agency, or a force of a *nature* different from the visible ones at the two ends of the line ? Is it not certain that the question is, How does the motion get from one to the other, whether there be a wire or not ? If there be a wire, it is plain that there is motion in it, for it is heated its whole length, and heat is known to be a mode of motion, and every molecule which is thus heated must have had some

antecedent motions. Whether it be defined or not,
and whether it be called by one name or another,
are quite immaterial, if one is concerned only with
the *nature* of the action, whether it be matter or
ether, or motion or abracadabra.

Once more : suppose we have a series of active
machines. (Fig. 11.) An arc lamp, radiating light-
waves, gets its energy from the wire which is heated,
which in turn gets its energy from the electric
current ; that from a dynamo, the dynamo from a

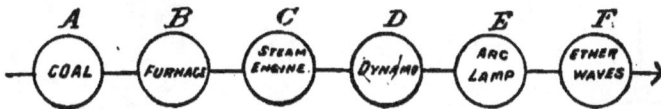

FIG. 11.

steam-engine ; that from a furnace and the chemical
actions going on in it. Let us call the chemical
actions A, the furnace B, the engine C, the dynamo
D, the electric lamp E, the ether waves F. (Fig. 12.)

The product of the chemical action of the coal is
molecular motion, called heat in the furnace. The
product of the heat is mechanical motion in the
engine. The product of the mechanical motion is
electricity in the dynamo. The product of the
electric current in the lamp is light-waves in the
ether. No one hesitates for an instant to speak of
the heat as being molecular motion, nor of the

motions of the engine as being mechanical; but
when we come to the product of the dynamo, which
we call electricity, behold, nearly every one says,
not that he does not know what it is, but that no
one knows! Does any one venture to say he
does not know what heat is, because he cannot
describe in detail just what goes on in a heated
body, as it might be described by one who saw with

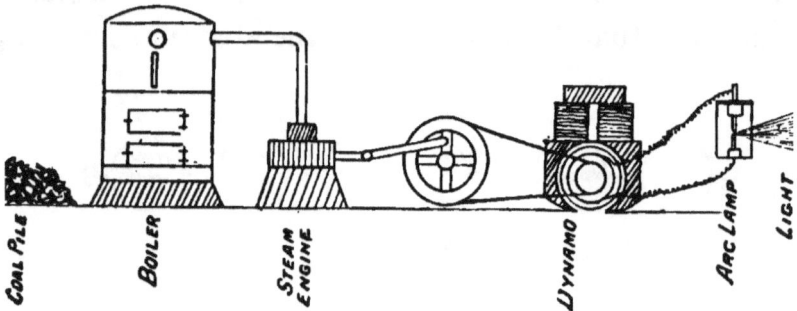

FIG. 12.

a microscope the movements of the molecules?
Let us go back for a moment to the proposition
stated early in this book, namely, that if any body
of any magnitude moves, it is because some other
body in motion and in contact with it has imparted
its motion by mechanical pressure.   Therefore, the
ether waves at F (Fig. 11) imply continuous motions
of some sort from A to F.   That they are all motions
of ordinary matter from A to E is obvious, because
continuous matter is essential for the maintenance

of the actions. At E the motions are handed over
to the ether, and they are radiated away as light-
waves.

A puzzling electrical phenomenon has been what

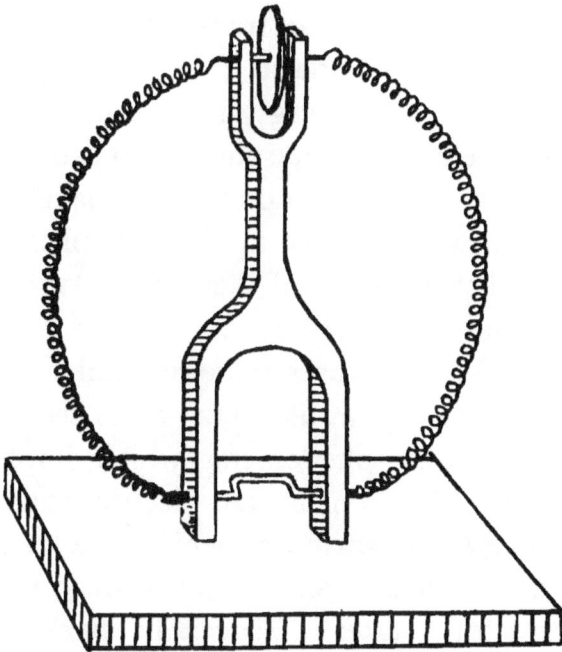

FIG. 13.

has been called its duality-states, which are spoken
of as positive and negative.   Thus, we speak of the
positive plate of a battery and the negative pole of
a dynamo ; and another troublesome condition to
idealize has been, how it could be that, in an
electric circuit, there could be as much energy at

the most remote part as at the source. But, if one
will take a limp rope, 8 or 10 feet long, tie its ends
together, and then begin to twist it at any point,
he will see the twist move in a right-handed spiral
on the one hand, and in a left-handed spiral on the
other, and each may be traced quite round the
circuit; so there will be as much twist, as much
motion, and as much energy in one part of the rope
as in any other; and if one chooses to call the right-
handed twist positive, and the left-handed twist
negative, he will have the mechanical phenomenon
of energy-distribution and the terminology, analo-
gous to what they are in an electric conductor.
(Fig. 13.) Are the cases more dissimilar than the
mechanical analogy would make them seem to be?

Are there any phenomena which imply that
rotation is going on in an electric conductor?
There are. An electric arc, which is a current in
the air, and is, therefore, less constrained than it is
in a conductor, rotates. Especially marked is this
when in front of the pole of a magnet; but the
rotation may be noticed in an ordinary arc by
looking at it with a stroboscopic disk, rotated so
as to make the light to the eye intermittent at
the rate of four or five hundred per second. A
ray of plane polarized light, parallel with a wire
conveying a current, has its plane of vibration
twisted to the right or left, as the current goes

one way or the other through the wire, and to a
degree that depends upon the distance it travels;
not only so, but if the ray be sent, by reflection,
back through the same field, it is twisted as much
more—a phenomenon which convinces one that
rotation is going on in the space through which
the ray travels. If the ether through which the
ray be sent were simply warped or in some static
stress, the ray, after reflection, would be brought
back to its original plane, which is not the case.
This rotation in the ether is produced by what is
going on in the wire. The ether waves called
light are interpreted to imply that molecules
originate them by their vibrations, and that there
are as many ether waves per second as of molecular vibrations per second. In like manner, the
implication is the same, that if there be rotations
in the ether they must be produced by molecular
rotation, and there must be as many rotations
per second in the ether as there are molecular
rotations that produce them. The space about a
wire carrying a current is often pictured as filled
with whorls indicating this motion (Fig. 14), and one
must picture to himself, not the wire as a whole
rotating, but each individual molecule inde-
pendently. But one is aware that the molecules
of a conductor are practically in contact with
each other, and that if one for any reason rotates,

the next one to it would, from frictional action, cause the one it touched to rotate in the opposite direction, whereas, the evidence goes to show that all rotation is in the same direction.

How can this be explained mechanically? Recall the kind of action that constitutes heat, that it is not translatory action in any degree, but vibratory, in the sense of a change of form of an elastic body, and this, too, of the atoms that make up the molecule of whatever sort. Each

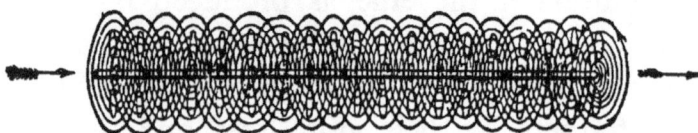

FIG. 14.

atom is so far independent of every other atom in the molecule that it can vibrate in this way, else it could not be heated. The greater the amplitude of vibration, the more free space to move in, and continuous contact of atoms is incompatible with the mechanics of heat. There must, therefore, be impact and freedom alternating with each other in all degrees in a heated body. If, in any way, the atoms themselves *were* made to rotate, their heat impacts not only would restrain the rotations, but the energy also of the rotation motion would increase the vibrations;

that is, the heat would be correspondingly
increased, which is what happens always when an
electric current is in a conductor. It appears
that the cooler a body is the less electric resistance
it has, and the indications are that at absolute zero
there is no resistance; that is, impacts do not
retard rotation, but it is also apparent that any
current sent through a conductor at that temperature would at once heat it. This is the same as
saying that an electric current could not be sent
through a conductor at absolute zero.

So far, mechanical conceptions are in accordance
with electrical phenomena, but there are several
others yet to be noted. Electrical phenomena has
been explained as molecular or atomic phenomena,
and there is one more in that category which is
well enough known, and which is so important and
suggestive, that the wonder is its significance has
not been seen by those who have sought to interpret
electrical phenomena. The reference is to the fact
that electricity cannot be transmitted through a
vacuum. An electric arc begins to spread out as
the density of the air decreases, and presently it
is extinguished. An induction spark that will
jump two or three feet in air cannot be made to bridge
the tenth of an inch in an ordinary vacuum. A
vacuum is a perfect non-conductor of electricity. Is
there more than one possible interpretation to this,

namely, that electricity is fundamentally a molecular and atomic phenomenon, and in the absence of molecules cannot exist? One may say, . "Electrical *action* is not hindered by a vacuum," which is true, but has quite another interpretation . than the implication that electricity is an ether phenomenon. The heat of the sun in some way gets to the earth, but what takes place in the ether is not heat-transmission. There is no heat in space, and no one is at liberty to say, or think, that there can be heat in the absence of matter.

When heat has been transformed into ether waves, it is no longer heat, call it by what name one will. Formerly, such waves were called heat-waves; no one, properly informed, does so now. In like manner, if electrical motions or conditions in matter be transformed, no matter how, it is no longer proper to speak of such transformed motions or conditions as electricity. Thus, if electrical energy be transformed into heat, no one thinks of speaking of the latter as electrical. If the electrical energy be transformed into mechanical of any sort, no one thinks of calling the latter electrical because of its antecedent. If electrical motions be transformed into ether actions of any kind, why should we continue to speak of the transformed motions or energy as being electrical? Electricity may be the ante-

cedent, in the same sense as the mechanical motion
of a bullet may be the antecedent of the heat
developed when the latter strikes the target; and
if it be granted that a vacuum is a perfect non-
conductor of electricity, then it is manifestly
improper to speak of any phenomenon in the
ether as an electrical phenomenon. It is from the
failure to make this distinction that most of the
trouble has come in thinking on this subject.
Some have given all their attention to what goes
on in matter, and have called that electricity; others
have given their attention to what goes on in the
ether, and have called that electricity, and some
have considered both as being the same thing, and
have been confounded.

Let us consider what is the relation between an
electrified body and the ether about it.

When a body is electrified, the latter at the same
time creates an ether stress about it, which is called
an electric field. The ether stress may be considered as a warp in the distribution of the energy
about the body (Fig. 15), by the new positions given
to the molecules by the process of electrification.
It has been already said that the evidence from
other sources is that atoms, rather than molecules,
in larger masses, are what affect the ether. One is
inclined to inquire for the evidence we have as to
the constitution of matter or of atoms. There is

only one hypothesis to-day that has any degree of
probability ; that is, the vortex-ring theory, which
describes an atom as being a vortex-ring of ether
in the ether. It possesses a definite amount of
energy in virtue of the motion which constitutes it,
and this motion differentiates it from the surrounding
ether, giving it dimensions, elasticity, momentum,
and the possibility of translatory, rotary, vibratory

NEUTRAL STRESS   POSITIVE STRESS   NEGATIVE STRESS

FIG. 15.

motions, and combinations of them. Without going
further into this, it is sufficient, for a mechanical
conception, that one should have so much in mind,
as it will vastly help in forming a mechanical conception of reactions between atoms and the ether.
An exchange of energy between such an atom and
the ether is not an exchange between different kinds
of things, but between different conditions of the
same thing. Next, it should be remembered that
all the elements are magnetic in some degree.

This means that they are themselves magnets, and every magnet has a magnetic field unlimited in extent, which can almost be regarded as a part of itself. If a magnet of any size be moved, its field is moved with it, and if in any way the magnetism be increased or diminished, the field changes correspondingly.

Assume a straight bar electro-magnet in circuit, so that a current can be made intermittent, say, once a second. When the circuit is closed and the magnet is made, the field at once is formed and travels outwards at the rate of 186,000 miles per second. When the current stops, the field adjacent is destroyed. Another closure develops the field again, which, like the other, travels outwards ; and so there may be formed a series of waves in the ether, each 186,000 miles long, with an electro-magnetic antecedent. If the circuit were closed ten times a second, the waves would be 18,600 miles long; if 186,000 times a second, they would be but one mile long. If 400 million of millions times a second, they would be but the forty-thousandth of an inch long, and would then affect the eye, and we should call them light-waves, but the latter would not differ from the first wave in any particular except in length. As it is proved that such electro-magnetic waves have all the characteristics of light, it follows that they must originate with electro-

magnetic action, that is, in the changing magnetism
of a magnetic body.   This makes it needful to
assume that the atoms which originate waves are
magnets, as they are experimentally found to be.
But how can a magnet, not subject to a varying
current, change its magnetic field ? The strength or
density of a magnetic field depends upon the form
of the magnet.   When the poles are near together,
the field is densest ; when the magnet is bent back
to a straight bar, the field is rarest or weakest, and
a change in the form of the magnet from a U-form
to a straight bar would result in a change of the
magnetic field within its greatest limits.   A few
turns of wire—as has been already said—wound
about the poles of an ordinary U-magnet, and
connected to an ordinary magnetic telephone, will
enable one, listening to the latter, to hear the pitch
of the former loudly reproduced when the magnet
is struck like a tuning-fork, so as to vibrate.   This
shows that the field of the magnet changes at the
same rate as the vibrations.

Assume that the magnet becomes smaller and
smaller until it is of the dimensions of an atom, say
for an approximation, the fifty-millionth of an inch.
It would still have its field ; it would still be elastic
and capable of vibration, but at an enormously
rapid rate ; but its vibration would change its field
in the same way, and so there would be formed

those waves in the ether, which, because they
are so short that they can affect the eye, we call
light. The mechanical conceptions are legitimate,
because based upon experiments having ranges
through nearly the whole gamut as waves in
ether.

The idea implies that every atom has what may
be loosely called an electro-magnetic grip upon the
whole of the ether, and any change in the former
brings some change in the latter.

Lastly, the phenomenon called induction may be
mechanically conceived.

It is well known that a current in a conductor
makes a magnet of the wire, and gives it an electro-
magnetic field, so that other magnets in its neighbourhood are twisted in a way tending to set them
at right angles to the wire. Also, if another wire
be adjacent to the first, an electric current having an
opposite direction is induced in it. Thus:

Consider a permanent magnet A (Fig. 15), free
to turn on an axis in the direction of the arrow. If
there be other free magnets, B and C, in line, they
will assume such positions that their similar poles
all point one way. Let A be twisted to a position
at right angles, then B will turn, but in the opposite
direction, and C in similar. That is, if A turn in
the direction of the hands of a clock, B and C will
turn in opposite directions. These are simply the

observed movements of large magnets. Imagine that
these magnets be reduced to atomic dimensions, yet
retaining their magnetic qualities, poles and fields.
Would they not evidently move in the same way
and for the same reason ? If it be true, that a
magnet field always so acts upon another as to tend
by rotation to set the latter into a certain position,

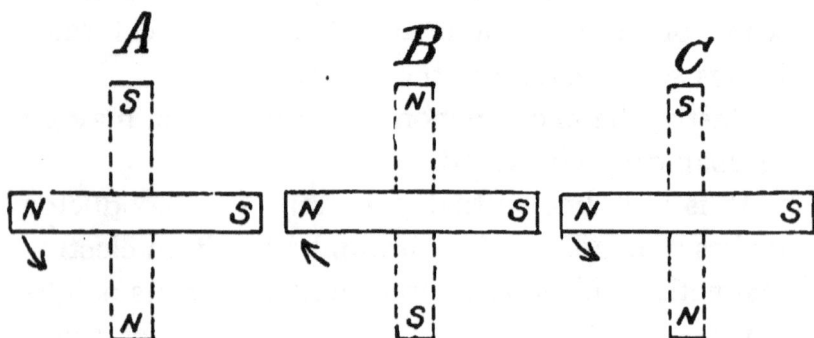

FIG. 16.

with reference to the stress in that field, then,
*wherever there is a changing magnetic field, there
the atoms are being adjusted by it.*

Suppose we have a line of magnetic needles free
to turn, hundreds or thousands of them, but dis-
arranged. Let a strong magnetic field be produced
at one end of the line. The field would be strongest
and best conducted along the magnet line, but every
magnet in the line would be compelled to rotate,
and if the first were kept rotating, the rotation

would be kept up along the whole line. This would be a mechanical illustration of how an electric current travels in a conductor. The rotations are of the atomic sort, and are at right angles to the direction of the conductor.

That which makes the magnets move is inductive magnetic ether stress, but the advancing motion represents mechanical energy of rotation, and it is this motion, with the resulting friction, which causes the heat in a conductor.

What is important to note is, that the action in the ether is not electric action, but more properly the result of electro-magnetic action. Whatever name be given to it, and however it comes about, there is no good reason for calling any kind of ether action electrical.

Electric action, like magnetic action, begins and ends in matter. It is subject to transformations into thermal and mechanical actions, also into ether stress—right-handed or left-handed—which, in turn, can similarly affect other matter, but with opposite polarities.

In his *Modern Views of Electricity*, Prof. O. J. Lodge warns us, quite rightly, that perhaps, after all, there is no such *thing* as electricity—that electrification and electric energy may be terms to be kept for convenience ; but if electricity as a term be held to imply a force, a fluid, an imponderable, or a thing which could be described by some one who knew enough, then it has no degree of probability, for spinning atomic magnets seem capable of developing all the electrical phenomena we meet.   It must be thought of as a *condition* and not as an entity.

THE END

www.ingramcontent.com/pod-product-compliance
Lightning Source LLC
Chambersburg PA
CBHW021939190326
41519CB00009B/1072